M000210993

SEMIOTEXT(E) FOREIGN AGENTS SERIES

© 2015 Franco "Bifo" Berardi

All rights reserved. No part of this book may be reproduced, stored in a retrieval system, or transmitted by any means, electronic, mechanical, photo-copying, recording, or otherwise, without prior permission of the publisher.

Published by Semiotext(e)
PO BOX 629, South Pasadena, CA 91031
www.semiotexte.com

Special thanks to John Ebert and Noura Wedell.

Cover art: Stuxnet No. 1 © James Hoff, 2014.

Design by Hedi El Kholti

ISBN: 978-1-58435-170-2
Distributed by The MIT Press, Cambridge, Mass. and London, England
Printed in the United States of America

AND

PHENOMENOLOGY OF THE END

SENSIBILITY AND CONNECTIVE MUTATION

Franco "Bifo" Berardi

to billino

Contents

Concatenation, Conjunction, and Connection

A rhizome has no beginning or end; it is always in the middle, between things, interbeing, *intermezzo*. The tree is filiation, but the rhizome is alliance, uniquely alliance. The tree imposes the verb "to be," but the fabric of the rhizome is the conjunction, "and ... and ... and ..." This conjunction carries enough force to shake and uproot the verb "to be." [... And to] establish a logic of the AND, overthrow ontology, do away with foundations, nullify endings and beginnings.

 —Gilles Deleuze and Félix Guattari, *A Thousand Plateaus*

The Metaphor of the Rhizome

In a rhizome there is no beginning and no end, according to Deleuze and Guattari, who propose that we view reality as an infinite rhizome, that is, an open concatenation of ands: and ... and ... and ...

This is why I'm writing this phenomenology of the end.

There is no end. Some may take this assertion as a source of endless hope; others may take it as a source of endless despair.

Both would be on the wrong path.

Do not get me wrong. I don't pretend to know what is good or bad. I am not hopeful, but neither am I hopeless. Phenomenology is an infinite task, so the phenomenology of the end must also be an interminable task.

I decided to stop writing this book here because my life is not endless, and I am approaching the end. But even so, I know that I will not stop concatenating: and, and, and.

In 1977, in the year of the premonition, Deleuze and Guattari wrote a short text called *Rhizome*, later published as the introduction to *A Thousand Plateaus*.

That year, social movements, punk culture and the dystopian imagination of art and literature foreshadowed in many ways a mutation that we are now witnessing and living through, and that has infiltrated the technological environment, social relations, and culture.

The rhizome is simultaneously the announcement of a transformation of reality, and the premise to a new methodology of thought. It is a description of the chaotic deterritorialization that follows modern rationalism, as well as a methodology for the critical description of deterritorialized capitalism.

This short text by Deleuze and Guattari foretold both the dissolution of the political order inherited from modernity, and the vanishing of the rational foundations of Western philosophy. At the same time, it opened the way to a new methodology that adopted what I call concatenation, rather than dialectical opposition, as a model to conceptualize cultural processes and social transformations.

Decades after the publication of this text, the rhizomatic metaphor can be seen as a way of mapping the neoliberal process of globalization, and the precarization of labor that it entails. But it also refers to the interminability of the philosophical task. But does the philosopher even have a task? And what, in that case, is that task? To map the territory of the mutation, and to forge conceptual tools for orientation in its ever-changing, deterritorializing territory: such are the tasks for the philosopher of our times.

Diachronic and Synchronic Phenomenology

A rhizomatic methodology shapes my approach to the subject of this book: the phenomenology of sensibility in our present age of technocultural mutation.

I argue that the ongoing transition from the alphabetical to the digital infosphere marks a shift from the cognitive model of conjunctive concatenation to a model of connective concatenation.

This book is concerned with the effects of this shift in the fields of aesthetic sensibility and emotional sensitivity.

The shift I am referring to is diachronic. It occurs as a transition, extending over a span of several human generations, during which time it transforms cognitive patterns, social behavior, and psychological expectations. But there is also a synchronic frame in which this shift occurs. Investigating that frame will allow me to describe the composition, conflicts, and coevolution of different psychocultural regimes as they simultaneously approach each other, collide, and interweave through the process of globalization.

The first, diachronic, and temporal axis of the phenomenology of sensibility that I am introducing here is the transition from the mechanical to the digital order, and the effects of this transition in the psychosphere.

The second, synchronic, and spatial axis of this phenomenology of sensibility is the coevolution of different cultural regimes of subjectivation in a globalized world.

During the last thirty years, the shift from the mechanical to the digital technosphere has provoked a mutation in the texture of human experience, and in the fabric of the world itself. The conjunctive mode of social interaction, which was prevalent since the Neolithic revolution, has been rapidly replaced by a connective mode of interaction. The latter began to prevail when the automating interfaces of the information machine pervaded and innervated the linguistic sphere.

I will try to describe the transition from the age of industrial capitalism to the age of semiocapitalism from the point of view of a shift from conjunction to connection as the dominant mode of social interaction.

Both sensibility and sensitivity are affected by this shift, although the mutation takes different forms and intensities in different geo-cultural areas of the world. I will thus trace the general lines of its aesthetic genealogy.

Sensibility will be my main concern: in these pages, I propose to draw a phenomenological map of the global mutation, investigating the aesthetic and the emotional side of sensibility.

For this purpose, I will trace the effects of the shift from the conjunctive to the connective mode in different geo-cultures.[1]

I must add that this research does not pretend to exhaustivity, as we know from Husserl that "phenomenology is an infinite task."

Sensibility and Creation

Emotion is a concatenation of unconnected things, events, and perceptions. But, we might ask, how is a concatenation possible between things that have no connection? Are there filters and grids that make the human organism sensitive to the color of autumn leaves, to the tenderness of a gesture, or to the sound of a song? Are the parts that enter into a concatenation fragments of a mosaic whose unity has been lost? Should we perhaps reconstruct the design to which the fragments once belonged? Or should we instead avoid presupposing a pre-existing design wherein segments are integrated and meaningful?

A conjunctive concatenation does not imply an original design that must be restored. A conjunction is a creative act; it creates an infinite number of constellations that do not follow the lines of a pre-conceived pattern, or an embedded program.

There is no design to fulfill at the beginning of the act of conjunction. Neither is there a model at the origin of the process of the emergence of form. Beauty does not correspond to a hidden harmony embedded in a universal spirit or in the mind of god. There is no code to comply with.

On the contrary, conjunctive concatenation is a source of singularity: it is an event, not a structure, and it is unrepeatable because it happens in a unique point in the network of space and time.

> The more we study the nature of time, the more we shall comprehend that duration means invention, the creation of forms, the continual elaboration of the absolutely new.[2]

According to Bergson, "we perceive duration as a stream against which we cannot go," a stream whose current we cannot move back up, and in this stream, new configurations of being arise out of nothing at every instant.

Sensibility is the faculty that makes it possible to find a path that does not yet exist, a link between things that have no intrinsic or logical implication. Sensibility is the sense-driven creation of conjunctions, and the ability to perceive the meaning of shapes once they have emerged from chaos. This does not happen by way of recognition, in the sense that such forms would be compatible with others that we would have seen before. It occurs because we perceive their aesthetic correspondence, their accordance, and conformity with the expectations of the conscious, sensitive, and sensible organism.

Expectations are crucial for the act of aesthetic conjunction, and for both the perception and the projection of forms. Such expectations are formed in the sphere of culture, which has a temporal history and a geographic location: what I call geo-cultures anchored in the flow of time. There is no implicit logic bringing together one sign with another, and their composition does not aim to arrive at an isomorphism with the world. The part is not completed through a conjunction with another part, nor do parts put side by side give life to a totality.

The only criterion of truth is the pleasure of the conjunction: you and I, this and that, the wasp and the orchid.

The conjunction is the pleasure of becoming other, and the adventure of knowledge is born out of that pleasure.

The problem is: How does it happen that under certain circumstances conjoined signs give birth to meaning? How does it happen that under certain circumstances conjoined events become history? And conjoined percepts become reality? Witold Gombrowicz suggests that reality is the effect of obsession.[3]

Gregory Bateson suggests that the skin is the line of conjunction and the sensible interface par excellence.[4] Forms are evoked and conjured within the aesthetic sphere. But what does *aesthetic* mean? By the word aesthetic, Bateson refers to everything that belongs to the sphere of sensibility. The latter is not the space where conjunction is recorded; instead, it is the factory of conjunctions. These do not happen somewhere in the world, they happen in a sensible mind.

For Bateson, the question of truth must shift from the realm of metaphysics and history to the realm of biology and sensibility. The mind is able to think life because it belongs to the living world. It's a matter of co-extensivity, not of representation. There is no ontological correspondence between the mind and the world, as the metaphysicians would like to believe. There is no historical totalization in which mind and world would coincide. There is no correspondence, adjustment, or *aufhebung*-realization. There are only conjunctions.

(And connections, as we'll see. But this is another story.)

Reality could be described as the point of conjunction of innumerable psycho-cognitive projections. If the mind can process the world as an infinite set of co-evolving realities that act on one another, this is only because the mind is in the world. Language is the realm where man brings forth being, and language is the conjunction of artificial fragments (signs) that produce a meaningful whole. But meaning does not take place in a preexisting nature or reality that exists as such, independently, it only occurs in the concatenation of minds.

Mirror Neurons, Language, and Connective Abstraction

When it comes to connection, the conceptual frame changes completely. When I use the word *connection*, I mean the logical and necessary implication, or inter-functionality, between segments. Connection does not belong to the kingdom of nature; it is a product of the logical mind, and of the logical technology of mind.

Since this text is essentially concerned with the anthropological and aesthetic effects of the shift from the sphere of conjunction to that of connection, I will return later to the distinction between conjunction and connection.

In his book *Saggio sulla negazione*, Paolo Virno argues that language, far from easing human contact, is, in reality, the basic source of conflict, misunderstanding, and violence.[5]

Only language establishes the possibility of negating what our senses are experiencing. Negation is like a switch that breaks the natural link between sensorial experience and its conscious elaboration. If immediate experience acknowledges a state of being, language can deny the state of being that is experienced. In this sense we can say that negation is the beginning of any mediation.

In the first pages of the book, Virno refers to biologist Vittorio Gallese's research on mirror neurons. According to Gallese and his colleagues, mirror neurons are what enable human beings to understand each other. They establish a net of inter-individual threads that trigger the process of understanding well before the individual becomes conscious of it. This implies that understanding is in fact a physical and affective phenomenon, before being an intellectual act.

According to Gallese, we understand the emotions and the actions of another person because, by looking at that person, we activate the same neurons that we would activate if we were feeling those same emotions, and performing those same actions.

We can call this mirror-like understanding *empathy*.

The development of linguistic competence, far from strengthening or confirming empathy, can be viewed as the beginning of the process of mediation that gradually erodes empathy, transforming understanding into a purely intellectual act of syntactic adaptation rather than a process of semantico-pragmatic osmosis.

According to Virno, language creates the un-natural possibility of reducing the light of immediate patency that surrounds perceptual experience. The order of language is syntactical: conventional rules open and close access to signification. In the course of human evolution, the syntactical order of language has invaded and re-framed the immediacy of empathy, and in many ways it has perverted or destroyed its very possibility.

In his book *Ah Pook Is Here*, William Burroughs conceived of language as a virus that spread as a mutation in the human environment.[6] Virno adds that the content of this virus is *negation*, a laceration in the canvas of the shared perceptions and projections that we call reality.

Empathy is the source of conjunction. Over the course of the history of civilization and of techno-evolution, it seems that the syntactization of the world, that is, the reduction of the common world to the syntax of linguistic exchange, has slowly eroded traces of empathic understanding, and instead, enhanced the space of syntactic conventions. Linguistic mediation has developed technologies that in turn shape the *umwelt*, that is, the surrounding environment.

With the digital, we have reached the end-point of this process of increasing abstraction, and an apex in the increasing dissociation of understanding from empathy.

In *Zero Degrees of Empathy*, the British psychologist Simon Baron-Cohen evokes *empathy erosion* to explain cruelty and violence between human beings. For Baron-Cohen, empathy consists of two causally-related steps: the first is the interpretation of the signs that proceed from the other, and thus the extrapolation of the other's feelings, desires, and emotions; the second is the ability to respond accordingly.[7]

I call conjunction this form of empathic comprehension. I call connection, on the other hand, the kind of understanding that is not based on an empathic interpretation of the meaningful signs and intentions coming from the other, but rather on compliance and adaptation to a syntactic structure. The best explanation of the difference between conjunction and connection occurs in the third book of Tolstoy's *War and Peace*, when Prince Andrey Bolkonski compares the game of chess with the game of war.[8]

The opposition between conjunction and connection is not a dialectical opposition. The body and the mind are not reducible in an oppositional way to either conjunction or connection. There is always some connective sensibility in a conjunctive body, and there is always some conjunctive sensibility in a human body formatted in connective conditions. It's a question of gradients, shades, and undertones, not one of antithetical opposition between poles.

Recomposition and A-Signifying Recombination

In the midst of infinite births and deaths, in the midst of decay, leaves falling from trees, and waves on the sea—all the infinite chaotic events that randomly occur in the universe—the only stunning and unexpected thing is our inexhaustible craving for sense, harmony, and order.

Metaphysical and dialectic philosophy focused on the idea of totality, a concept that was based on the assumption either of a pre-existing order, or of a final order that it would restore or bring into being. According to the principles of totalitarian philosophy, each fragment would find its pre-established place, and all parts were arranged so as to compose an original or final totality, code, or destiny.

The phenomenological approach takes leave of the assumption that knowledge can lead to the perfect totality, and abandons the project of a totalitarian identification of thought and world. It thus opens the way to the possibility of different theoretical constructions, based on different *erlebnisses*, or forms of life. A rhizomatic methodology is just one among a multiplicity of possible phenomenological approaches.

According to a rhizomatic methodology, meaning emerges from a vibration that is singular in its genealogy, and can proliferate and be shared. Meaning is therefore an event, not a necessity—and we can share it with other singularities that enter into vibrational syntony, or sympathy, with our meaningful intentions.

A rhizomatic methodology does not presuppose or imply any totality that it would establish or restore. It is based on the principle of non-necessary conjunctions, and on the continuous molecular recomposition of cells, to borrow from a scientific vocabulary, whose destination is not implied in their program or genetic code.

Recomposition is a process of uncertain and autonomous subjectivation, where flows of enunciation interweave and create a common space of subjectivity. This collective subjectivity can be the result of an imagined form of belonging, such as a tribe, a nation, or a common faith. In this kind of collective existence, enunciation pretends to bring about truth, and divergence is seen as betrayal.

But collective subjectivity can also be the expression of an attraction: for example, desire as the singular creation of the other as singularity. In this case we can speak of a collective singularity, a singularity that is the living experience of a pathway from nowhere to nowhere. As Antonio Machado writes, and the Zapatistas repeat: *"Caminante no hay camino, el camino se hace al andar."*

In this case, desire, as an attraction to singularity, generates the pathway and is the reason for collective existence (its *raison d'être*).

Rather than the homeland, the family, or ideological dogma, the collective subjectivity that I am trying to trace here is based on nomadic desire, not on belonging, or code.

I use the term recomposition to describe this process of social conjunction, that is, the opening and conjoining of individuals into a collective singularity, through which they express an affective and political solidarity that does not rely on identification, conventional codes, or marks of belonging.

Recomposition is the meeting, converging, and conjoining of singular bodies on a path that they share, provisionally, for a time. That common path is not inscribed in a genetic code, or in a cultural belonging—rather, it is the discovery of a common possibility that is the meeting point of singular drifts of desire. The community that results from the process of recomposition is a community of desire, not one of necessity. This is quite different from the process of recombination, where a-signifying segments are connected in accordance with coded rules of generation.

Conjunction versus Connection: The Ongoing Mutation

I call conjunction a concatenation of bodies and machines that can generate meaning without following a pre-ordained design, and without obeying any inner law or finality.

Connection, on the other hand, is a concatenation of bodies and machines that can only generate meaning by following an intrinsic, human-generated design through obeying precise rules of behavior and functioning.

Connection is not singular, intentional, or vibrational. Rather, it is an operative concatenation between previously formatted agents of meaning (bodies or machines) that have been codified, or formatted according to a code.

Connection generates messages whose meaning can only be deciphered by an agent (a body, a machine) that shares the same syntactic code that generated the message.

In the sphere of conjunction, the agent of meaning is a vibrating organism, where vibration refers to the uncertain and unresolved oscillation around an asymptotic point of isomorphism.

The production of meaning is the effect of the singularization of a series of signs (traces, memories, images, or words…).

Conjunction is the provisional and precarious syntony of vibratory organisms that exchange meaning.

The exchange of meaning is based on sympathy, the sharing of *pathos*.

Conjunction, therefore, can be viewed as a way of becoming other. Singularities change when they conjoin, they become something other than what they were before, in the same way as love changes the lover, or the conjunctive composition of a-signifying signs gives rise to the emergence of previously inexistent meaning.

By contrast, in the connective mode of concatenation, each element remains distinct and only interacts in a functional way. Rather than a fusion of segments, connection entails a simple effect of machine functionality.

In order for the connection to be possible, segments must be linguistically compatible. Connection thus presupposes a process whereby the elements that need to connect are rendered compatible.

The digital web, for example, extends through the progressive reduction of an increasing number of elements into a format, a standard, and a code that renders different elements compatible.

The considerations above are meant to introduce what I take to be the anthropological mutation that is underway in our times, essentially, a transition from the predominance of the conjunctive mode to the predominance of the connective mode in the sphere of human communication.

From the anthropological point of view, this technocultural change is centered on the shift from conjunction to connection in the paradigms of exchange between conscious organisms.

The leading factor of this change is the insertion of electronic segments into the organic continuum, the proliferation of digital devices in the organic universe of communication and in the body itself.

This leads to a transformed relation between consciousness and sensibility, and an increasingly desensitized exchange of signs.

Conjunction is the meeting and fusion of round or irregular bodies that are continuously weaseling their way about without precision, repetition, or perfection. Connection is the punctual and repeatable interaction of algorithmic functions, straight lines and points that overlap perfectly, and that plug in or out according to discrete modes of interaction that render the different parts compatible to a pre-established standard.

Passing from conjunction to connection as the predominant mode of conscious interaction between organisms is a consequence of the digitalization of signs, and of increasingly mediatized relations.

This digitization of communicative processes induces a desensitization to the curve, and to the continuous process of slow becoming, along with a concurrent sensitization to the code, or to sudden changes of state.

Conjunction entails a semantic criterion of interpretation. In order to enter into conjunction with another organism, the first organism sends signs to the other, signs whose meaning can only be interpreted in the pragmatic context of their interaction by tracing an intention, a shade of what remains unsaid, conscious and unconscious implications, and so on.

Connection instead requires a purely syntactic criterion of interpretation. The interpreter must recognize a sequence and be able to carry out the operation that is foreseen by the *general syntax* (or operating system); there is no margin for ambiguity in the exchange of messages, nor can intention be manifest though nuances.

The process of this gradual translation of semantic interpretations into syntactic differences runs from modern scientific rationalism, to cybernetics, and artificial intelligence programs.

Connective Logic

The debate on artificial intelligence began in the 1960s.

To outline the problem that lies at the core of artificial intelligence, Hubert Dreyfus distinguished between "areas in which relevance has been decided beforehand [...], and areas in which determining what is relevant is precisely the problem."[9]

When we exchange messages in the conjunctive sphere, we are trying to find out what is relevant for those who are participating in the communication. We don't know what our common object of interest and attention is: communication is about shedding light on that point. In the connective sphere, on the contrary, we start from a common ground of conventional knowledge, translated into technological standards and formats that make connection possible.

Concerning the genesis of connective methodology in the history of modern philosophy, Hubert Dreyfus writes:

> As Galileo discovered that one could find a pure formalism for describing physical motion by ignoring secondary qualities and teleological considerations, so, one might suppose, a Galileo of human behavior might succeed in reducing all semantic considerations (appeal to meanings) to the techniques of syntactic (formal) manipulation.
>
> The belief that such a total formalization of knowledge must be possible soon came to dominate Western thought. [...] Hobbes was the first to make explicit the syntactic conception of thought as calculation. [...] Leibniz thought he had found a universal and exact system of notation, an algebra, a symbolic language, a "universal characteristic" by means of which "we can assign to every object its determined characteristic number."[10]

Dreyfus then retraces the steps that led to the formation of the contemporary digital mind-set.

> An important feature of Babbage's machine was that it was digital. There are two fundamental types of computing machines: analogue and digital. Analogue computers do not compute in the strict sense of the word. They operate by measuring the magnitude of physical quantities. Using physical quantities such as voltage, duration, angle of rotation of a disk, and so forth, proportional to the quantity to be manipulated, they combine these quantities in a physical way and *measure* the results. A digital

computer [...] represents all quantities by discrete states, for example, relays which are open or closed, a dial which can assume any one of ten positions and so on, and then literally *counts* in order to get results. [...] since a digital computer operates with abstract symbols which can stand for anything, and logical operations which can relate anything to anything, any digital computer [...] is a universal machine.[11]

The universal digital machine is the logical and technological condition of our contemporary anthropological mutation.

Conjunction is the opening of bodies to the understanding of signs and events, and their ability to form organic rhizomes, that is, concrete, carnal concatenations of pulsating vibratory bodily fragments with other pulsating vibratory bodily fragments.

On the contrary, in a digital environment, only what fulfills the standard of compatibility can connect, meaning that certain elements will be unable to connect to others. In order for distant communicative agents to be able to connect, we must provide them with tools enabling them to access the flow of digital information.

When connection replaces conjunction in the process of communication between living and conscious organisms, a mutation takes place in the field of sensibility, emotion, and affect.

As I have noted before, this mutation occurs in time, in the diachronic dimension of the transition from the modern mechanical environment of indust-reality to the postmodern environment of semio-economy. But it is not homogeneous, as it depends on the particular features of the cultural context, geo-cultural and synchronic, in which it takes place.

I will thus turn to selected, synchronic cultural contexts to investigate the different forms of this diachronic, connective mutation, with a special attention to the relation between aesthetic sensibility and forms of emotional life.

Evolution and Sensibility

The expression *cognitive wiring* refers to the capture and submission of life and mental activity into the sphere of calculation. This capture occurs on two different levels: on the epistemic level it implies the formatting of mental activity, on the biological one it implies the technical transformation of the processes by which life is generated.

In the modern age, the modeling of the body was essentially macro-social and anatomical—as Michel Foucault has extensively shown in his works about the genealogy of modernity. The subjection of the social body to industrial discipline was linked to the macro-social action of repressive machines acting on the individual body.

Today, digital technology is based on the insertion of neuro-linguistic memes and automatic devices into the sphere of cognition, into the social psyche, and into forms of life. Both metaphorically and non-metaphorically, we can say that the social brain is undergoing a process of wiring, mediated by immaterial linguistic protocols as well as by electronic devices.

As generative algorithms become crucial in the formation of the social body, the construction of social power shifts from the political level of consciousness and will to the technical level of automatisms located within the process of generating linguistic exchange and forming psychic and organic bodies.

My attention here will be focused on the biosocial modeling of sensibility, that is, on the embedding of cognitive automatisms at the deep levels of perception, imagination, and desire. This implies that social becoming is no longer understandable in the framework of history but in the framework of evolution.

History is the conceptual sphere where conscious voluntary actors transform the conditions and social structures that surround them. In the sphere of evolution, on the other hand, human beings cannot be considered actors because evolution refers to the natural becoming of organisms in their interaction with the environment.

From the point of view of intentionality, the concepts of history and evolution can be distinguished, and opposed. The concept of history, emphasized by the romantic tradition, was particularly important to the Hegelian dialectical tradition, including Marx and the Marxist movement. The concept of evolution, on the other hand, was elaborated in a cultural space more akin to the positivist school of thought.

Historical action takes place when political intentionality is effective in modeling the environment. Evolution, on the contrary, occurs when the exchange between humans and nature, and the reciprocal transformation of these terms cannot be controlled by intentional political action.

In today's conditions of hyper-complexity and technological acceleration, the social sphere can no longer be properly under-stood in terms of political transformation. It is better explained through evolution, particularly neural evolution. Indeed, the evo-lution of the brain resulting from environmental action on cognition and society, and the subjective adaptation of the human mind are today the main factors of social transformation, and can hardly be subjected to political will.

In the context of history as outlined above, political action was driven by will, rational understanding, and prediction—while in the context of evolution, the organism is understood to become attuned to its environment, with sensibility being the faculty that makes this syntonization possible. Consequently, the relevance and effectiveness of human action no longer occurs at the level of rational knowledge, political decision, and will, but instead at the level of intuition, imagination, and sensibility.

Clearly, the conceptual and practical sphere of modern politics has lost its ground.

In the age that began with Machiavelli and culminated with Lenin, human will (the prince, the state, the party) was able to reign on the infinite chaotic variation of events and projects, and subject individual interests and passions to the common goal of social order, economic growth, and civil progress.

The technical transformation that we witnessed in the last decades of the twentieth century, the infinite proliferation of sources and flows of information, unleashed by the acceleration of network technology, has rendered impossible the conscious elaboration of information by the individual mind, and the conscious coordination of willful individual agents.

The loss of effectiveness of political action is essentially due to a change in temporality: with the acceleration and complexification of the infosphere, reason and will—those essential tools of political action—can no longer process and decide in time. Technical transformation has radically altered the conditions of mental activity and the forms of interaction between the individual and the collective spheres.

In the age of voluntary action that was called modernity, these two spheres—the individual and the collective—could be seen as distinct, externally linked, and interacting on the basis of an effective intentionality.

Today, the distinction between the individual and the collective has been blurred. Crowds and multitudes are involved in automatic chains of behavior, driven by techno-linguistic dispositives. The automation of individual behavior—since individuals have been integrally penetrated and concatenated by techno-linguistic interfaces—results in a swarm effect. If the human is the animal who shapes the environment that shapes his/her own brain, the swarm effect is thus the outcome of the human transformation of its technical environment, leading to the subjugation of mental behavior.

PART 1

SENSIBILITY

1

The Sensitive Infosphere

The human skin of things,
the dermis of reality [...]
—Antonin Artaud, *Selected Writings*

The Sensible Organism

What do we mean by sensibility? In his book on Francis Bacon, Deleuze writes that:

> Sensibility is vibration. We know that the egg reveals just this state
> of the body "before" organic representation: axes and vectors, gra-
> dients, zones, cinematic movements, and dynamic tendencies, in
> relation to which forms are contingent or accessory.[12]

Sensation is the opening to the world that enables the tantric egg of the body without organs to vibrate.

> A [Body without organs] is made in such a way that it can be
> occupied, populated only by intensities. Only intensities pass and
> circulate. Still, the BwO is not a scene, a place, or even a support

upon which something comes to pass. It has nothing to do with phantasy, there is nothing to interpret. The BwO causes intensities to pass; it produces and distributes them in a *spatium* that is itself intensive, lacking extension. It is not space, nor is it in space; it is matter that occupies space to a given degree—to the degree corresponding to the intensities produced. It is nonstratified, unformed, intense matter, the matrix of intensity, intensity = 0; but there is nothing negative about that zero, there are no negative or opposite intensities. Matter equals energy. Production of the real as an intensive magnitude starting at zero. That is why we treat the BwO as the full egg before the extension of the organism and the organization of the organs, before the formation of the strata; as the intense egg defined by axes and vectors, gradients and thresholds, by dynamic tendencies involving energy transformation and kinematic movements involving group displacement, by migrations: all independent of *accessory forms* because the organs appear and function here only as pure intensities. The organ changes when it crosses a threshold, when it changes gradient. "No organ is constant as regards either function or position, ... sex organs sprout anywhere, ... rectums open, defecate and close, ... the entire organism changes color and consistency in split-second adjustments." The tantric egg.[13]

The organism is the specific sedimentation of the vibrations through which the potentiality of the egg is actualized, and it naturally retains the ability to return to the state of a body without organs whenever it finds the potential to change again. An organism is doomed to lose sensibility and to lose its ability to vibrate when it stiffens its obsessions, its codes of interpretation, and what Deleuze and Guattari call its *ritornello*, or refrain.[14]

Sensibility can be defined as the faculty that enables the organism to process signs and semiotic stimulations that cannot be verbalized or verbally coded. Someone who is unable to comprehend moods, emotions, allusions and the unsaid—a large part of what constitutes communication and daily affective and social life—is commonly said to lack sensibility.

Like a thin film recording and deciphering non-verbal impressions, sensibility allows human beings to join together and connect through relations of empathy, or in other words, to *regress* to a non-specified and non-codified state of bodies without organs that pulsate in unison. It is through empathic relations that we can understand signs that are irreducible to information, and yet that constitute the foundation of inter-human understanding. Sensibility is the faculty of decoding intensity, which by definition means to escape the extensive dimension of verbal language.

Sensibility is the ability to understand what is unspoken.

Beauty

According to Paul Klee, the task of creative activity is not to reproduce the visible, but to *make* visible. Sensibility is the faculty of making visible a configuration of the world.

Poetry, music, painting, cinema, literature do not simply represent existing reality, their function is to make the world sensibly perceivable, to translate the world into sensitive configurations. Although art and sensibility are not limited to the realm of beauty, we will call beauty the emergence of forms in the realm of sensibility. In this sense, it is not necessarily linked to notions of harmony. Indeed, there can be beauty in symmetry or in the harmony intrinsic to an object, but a violation of symmetric order can cause no less aesthetic pleasure.

Beauty implies a regression to the state of a body without organs, where it is possible to create new constellations of meaning and new functionalities for the objects of experience.

The Russian literary theorist Victor Shklovsky defines poetry as the restitution of *pathos* to words that have been overly used and consumed: he calls *ostranenie* (estrangement) the procedure that returns meaning and energy to such signs. Rather than looking to symmetry and dissymmetry, Shklovsky invokes estrangement to explain aesthetic emotion and the pleasure of forms, that is, an unpredicted deviation in the relationship between sign and meaning.

This is the point: beauty occurs with the derailing of the customary, predictable relation between sign and meaning, and with the discovery of unforeseen and multiple perspectives. Beauty has to do with surprise. Symmetry and dissymmetry are simply modalities of the configurations of signs, whereas aesthetic value is dependent on improbability, unpredictability, and strangeness: a distance from the predictable order.

Perhaps beauty is an ironic tolerance for the imperfections of real life, relaxing the tensions between an organism and its environment, between mind and body, between existence and being for death.

But it may also be something entirely different, the cruelty of the inexhaustible, for one. And indeed, life continuously produces bodies that we cannot enjoy; they pass us by unshaken, indifferently brushing against our gaze. Beauty is the cruelty of this infinite excess of nature. It is the sudden awareness of the fragility of our conscious organism, the intuition of the impossible infinity of experience.

Art, aesthetic creation, and the gesture of suspension play with this tolerant irony of beauty. But there is also the beauty that the language of eroticism de-reasons about, that gushes out of the blind and merciless game of nature, in the tenderness of sensual energy and decomposition.

The Western philosophical tradition has conceived of aesthetics as a theory of beauty. But this conception has revealed itself to be rather inconclusive. I think it would be better to conceive of aesthetics as the science of semiotic emanation in its interaction with sensibility. Aesthetics should return to its etymon. In that sense, rather than to beauty (which is a quality of the object in itself), it would refer to sensibility, the experience of the object.

Democritus conceived of sensation as the chemical integration of sensuous and environmental matter. Sensibility can be seen as a modulation of this *syntonia* (tuning). Sensibility is the certainty of judgment, the singular certainty of what is good or bad for an organism. According to Gabrielle Dufour-Kowalska,

> Sensibility is not simply the faculty of the beautiful, and artistic beauty does not constitute a separate realm. [...] Sensibility belongs to a sphere of certainty that no objective knowledge can lay claim to, because the real source of human knowledge is not the intellect, but sensibility. [...] In its radical subjectivity, sensibility is the faculty of the real.[15]

What causes harmonious or disharmonious vibrations between a singularity and the cosmos? What are the similarities and differences between aesthetic and erotic pleasure? Should we postulate a neurophysiological predisposition in humanity, an innate program for the sensuous reception of the world, in other words, a bio-grammar of aesthesia and eroticism? Or should we suppose instead that the conditions of harmony are exclusively cultural?

Certainly, harmony and disharmony are not intrinsic to the cosmos. They are modalities of a relationship between the singular receptive psyche and the becoming of the cosmos: here lies the secret of pleasure and beauty.

Genealogy of the Skin

The scene this book is investigating stretches between the two poles of techno-semiotic emanation and sensibility. The cultural, historical, and social becoming of the planet can be viewed according to the techno-semiotic and cultural modulations of human sensibility.

Accelerated by the power of technologies, the environment now exceeds any possibility of human measure. Think of the hyper-saturation of the mediascape that is overwhelming the capacity for critical thought. Human reason is exhausted. The infinite complexity of phenomena overwhelms our capacity for observation. Thrust beyond the realm of the properly human through its interface with technology, sensibility has now incorporated the inorganic.

Sensibility can be regarded as a particular realm of what Foucault defines as the episteme: the shaping of social perception that enforces a unitary projection of the world, and leads to social discipline.[16] Semiocapitalism has deeply penetrated the neural circuits of social culture through technology's permeation of sensibility. Let us now outline a phenomenology of the mutation of sensibility.

First, we must distinguish the sensory from the sensuous. The sensory is the perceiving faculty of the organism, while the sensuous is the selective attraction/repulsion that the organism projects onto its surrounding environment. Sensibility is the singular faculty that allows for a projection of the real. As such, it is morphogenetic, and continuously creates forms. Sensibility is the certainty of judgment in this respect, because aesthetic judgment does not apply to what is separate from pleasure or pain, and thus entails the singular certainty of good and bad.

Here, the idealist vision is turned upside down. Whereas for Hegel, art, that is, the activity of morphogenetic creation, was a moment of the process towards knowledge, we would say on the contrary that thought is a moment of sensibility, the tuning of *atman* and *prana*, singular breath and cosmic rhythm.

If thought tends towards the conceptual capture of the world, sensibility caresses and shapes the world without interrupting its becoming, without claiming to establish any absolute truth.

The epidermis is the point of contact, the sensitive interface between the conscious self and the infinite emission of signs. In the night sky, desire is the order of constellations. The epidermis is the stratum where order is cracked opened and created along the coordinates of pleasure and pain. Among the infinite signs coming from the cosmos and from the artificial infosphere, constellations emerge that are ruled and designed by epidermal intuition and desire as they create, compose, choose, hide and make worlds. But the epidermis is not a biological or natural stratum. The skin is shaped by touch, caresses, suffering and scars. The infosphere shapes the sensors that create world constellations in the infosphere. The epidermis is a memory of caresses. It is the interface of the social, and its sensibility is the place of the utmost intensity of the mutation underway.

I want to question here the common assumption that sensibility is passive and purely receptive, while imagination implies creation, falsification, and simulation.

I view sensibility as action on the environment and emanation, as well as reception and perception.

The Infosphere and Sensibility

The infosphere is the sphere of the intentional signs that surround the sensible organism.

Perception and the technological architecture surrounding the perceptive organism are intertwined. Marshall McLuhan's theoretical innovations consisted essentially in that breakthrough, understanding that the technical structure of semiosis, the format of emission of semiotic flows, shapes perception and imagination.

In the Renaissance, man's perception of the space of everyday life changed because of innovations in representational technique, in particular, through the invention of perspective.

Prior to modernity, the infosphere was characterized by a regime of slow transmission, and this slow pace shaped lived time and cultural expectations. Throughout the history of civilization, perception has always been molded by artificial regimes of images, and by techniques to produce and circulate representations of the world. The modern acceleration of the transmission of signs and the proliferation of sources of information has transformed the perception of time. The infosphere has grown more rapid and dense, and the proliferation of info-stimuli has subjected sensibility to mutagenic stress. Due to an intensification of electronic signals, the acceleration of the infosphere is dragging sensibility into the vertigo of simulated stimulation. This leads to a reshaping of the perception of the other and of its body. Pressure, acceleration, and automation are affecting gestures, postures, and the whole of social proxemics.

As images proliferate, our faculty of imagination undergoes a vertiginous acceleration. The image is not the brute perception of empirical data brought to our attention by matter, but the imaginational elaboration of visual matter by our mind, and the technical mode in which we receive and elaborate images acts upon the formation of our imagination.

Techno-media adjustments and psycho-cognitive mutations are as interdependent as the organism and its ecosystem. The conscious organism is also a sensuous organism, since it is a bundle of sensitive receptors. The connective technosphere we inhabit today resembles the outcome of a projective zapping where we combine sequences coming from different sources. The social unconscious reacts to this continuous deterritorialization in various ways, through adaptation, disconnection, or pathology.

Suddenly awoken by the eruption of semiotic proliferation and deprived of the filters inscribed in the critical and disciplinary mindset of modernity, the nucleus of identity is fleeing and dissolving in all directions.

Emotion in Cybertime

Let's call the *infosphere* the universe of transmitters, and the *social brain* the universe of receivers. The universe of receivers—human beings made of flesh, and frail and sensuous organs—is not formatted according to the standards of digital transmitters. Although the neural system is highly plastic, and can mutate according to the rhythm of the infosphere, the format of the transmitter does not correspond to the format of the receiver. So what happens? As the electronic universe of transmission interfaces with the organic world of reception, it is producing pathological effects: panic, over-excitement, hyperactivity, attention deficit disorders, dyslexia, information overload, and the saturation of neural circuitry.

In late modernity, with the transition from the alphabetical to the electronic regime of communication, the universe of transmission has been undergoing constant acceleration, while the universe of receivers has desperately tried to follow its rhythm, accelerating and standardizing cognitive response.

Although the neural system is plastic, the human mind evolves with a rhythm that is totally different from the rhythm at which machines evolve. This is why the expansion of cyberspace implies an acceleration of cybertime that has pathological effects on the living terminal, the human mind with its physical, emotional, and cultural limits.

Multitasking implies the quick shift from one informational frame to another. Although the human mind seems perfectly suited for multitasking, doing so actually triggers a psychological mutation, producing new forms of mental suffering such as panic, attention deficit disorders, burnout, mental exhaustion, and depression.

We are caught in a frenzy of enforced socialization. To produce and to work implies that we be connected; connection equals work. Economic obsession brings about a permanent mobilization of productive energy. According to Jonathan Crary, "this is the form of contemporary progress—the relentless capture and control of time and experience."[17]

This is the main focus of the semio-corporation. Its mission is to render the relation between the Net and the netter, between the machine and the cognitive worker flexible and dynamic: Google.

The overproduction that led to cyclical crises in the age of industrial capitalism, according to Karl Marx, becomes permanent in the sphere of semiocapitalism, as the proliferation of sources of nervous stimulation implies the infinite overload of the attention market. According to Crary, the expansion of demands on attention leads to the permanent siege, and the relentless expansion of the time we are required to be alert. Attention has become the scarcest of resources: since we no longer have time for conscious attention, we must deal with information and make decisions in an increasingly automatic way. We thus tend to be governed by decisions that do not respond to long-term rational strategies, but rather to simple binary alternatives.

The psychiatrist Eugene Minkowski, author of the book *Lived Time: Phenomenological and Psychopathological Studies* published in 1933, stressed the link between mental suffering and the perception of time, that is, the way we perceive how we flow through time, or whether our moods of experiencing life are lazy or frantic.

Clearly influenced by Henri Bergson, for whom time is duration or the projection of an existential *vécu*, Minkowski does not speak of *time*, but of *lived time*. Following Minkowski's lead, the prevailing forms of contemporary psychopathology might fall under the label of chronopathology.

In the American school system, the diagnosis of ADHD (Attention Deficit Hyperactivity Disorder) is increasingly common. This disorder manifests as hyperactivity, and the resultant inability to focus attention on a subject for sustained periods of time. It seems that daily exposure to electronic flows of psycho-stimulation from an early age leads to changes in affective and emotional makeup, language, imagination, and the very perception of lived time.

Through the process of work, humans are transformed into connected elaborators of information. Since increases in productivity are based on the acceleration of information flows, the temporal contraction and the acceleration of brain activation that this implies has the effect of weakening personal experience.

While *cyberspace*, the virtual dimension of info-productive interaction between agents of communication, can be infinitely expanded, *cybertime*, that is, the duration of perception, cannot be expanded beyond certain limits for it is bounded by emotional and cultural temporality, as well as by organic restrictions. The emotional and cultural elaboration of stimuli occurs in time, and the time necessary for psychological and bodily elaboration cannot be shortened beyond a certain point.

As the amount of information demanding our attention expands, there is less and less attention time available for elaboration. The technical composition of the world has changed, but the modalities of cognitive appropriation and elaboration cannot adapt to this change in a linear way. The technical environment is changing at a much faster rate than culture, and than cognitive behavior in particular.

Although we can extend our exposure to information flows, and we can increase our efficiency through drugs, experience cannot be intensified beyond a set limit. Acceleration is in fact impoverishing experience, since the intensive modalities of pleasure and knowledge have become stressed to the point of exhaustion.

This conflict, or incompatibility, between cyberspace and cyber-time is a striking paradox of our society, and because of capitalist exploitation it is producing pathological effects.

It also leads to diminished sensibility. Beyond a certain limit, the experience of acceleration leads to a contraction of the time available for conscious elaboration and thus to a loss of sensibility—which also has ethical consequences. Sensibility exists in time, and cyberspace has grown so thick that the sensible organism—as a conscious singularity—no longer has the time to extract meaning and pleasure from experience.

Drugs for erectile dysfunction such as Viagra and similar products have more to do with attention than with physical impotence. Since the time for caresses and conversation is no longer available for precarious lovers, fast sex requires pharmaceutical support. It is sex without attention, since attention requires time.

Prozac-Crash

Today, the universe of transmitters, cyberspace, can no longer be translated by the universe of receivers, cybertime. Here lies a pathogenic gap, which the flourishing psychopharmaceutical industry occupies by selling more pills every year, since drugs have become the only way to manage mental suffering, anxiety, and sadness.

At the end of the 1970s, when workers were forced to accelerate productivity, a huge epidemic of drug addiction swept the late-industrial metropolitan areas. As capitalism entered the age of neo-human acceleration, cocaine, a substance that accelerates mental and bodily rhythms, became very fashionable. In the same period, however, many people started injecting heroin, a substance that deactivates the link between individual perception and the perception of the rhythm surrounding the person. From the 1970s to the 1980s, such white powder epidemics triggered both existential and cultural devastations whose traces can be found in the music, literature, and visual arts of the American no-wave and of the British punk culture. The use of psychopharmaceutical products continued to spread, and we entered the age of antidepressants and mood enhancers, Prozac, Xanax, Zoloft, and so on.

As semiocapitalism is based on the constant exploitation of mental energy, and competition constitutes the general form of relations in the precarious labor market, mental suffering has become a social epidemic. The main source of pathology is competition in the area of interpersonal relations. The individual symptoms of this epidemic are the constant stress on attention, the reduction of the time available for affection, loneliness, existential misery, then angst, panic, and depression. In this way, psychopathology and economics are increasingly linked. In the transition to semiocapitalism, mental suffering no longer concerns a small minority of weird people, but tends to become the norm in a system that is based on the exploitation of precarious cognitive work.

As long as capitalism was seeking to extract physical energy from the bodies of salaried people, psychopathology could remain secluded in marginal spaces. Who cared about your suffering, as long as you were screwing, hammering, working on a lathe, and assembling the parts of a machine? You could feel as alone as a fly in a bottle, but your productivity wasn't hindered by your loneliness and pain, since your muscles could still work. Today, on the contrary, semiocapitalism requires neural energy for mental work, and alienation is exploding at the core of the social machinery. Ups and downs, panic and depression are words shared between economic parlance and psychopathology. These terms are not metaphorical; they are clues to the growing interdependence of economic behavior and mental pathology.

After the end of the avant-garde, after the infiltration of art into the territory of social communication, aesthetic stimulation spread aggressively to every part of the mediascape, into advertising, television, design, and web design. The conscious and sensible organism is enveloped by a semiotic flow that is not only an information carrier, but also a factor of perceptual stimulation and psychological excitement. Widespread aestheticization absorbs erotic energy, and diverts it from the body towards signs.

Classical aesthetic philosophy was based on the conceptual and sensible centrality of catharsis. In the Aristotelian model, art was conceived as the trigger of a captivating wave that provoked excitement and led to a climax, a cathartic emotion. In the classical, the romantic and the modern conceptions, beauty was identified with the moment of culmination that relaxed the tension implied in the relation between the sensitive body and the world: catharsis, harmony, and sublime detachment. Reaching the cathartic climax of aesthetic emotion is an event that can be assimilated to the orgasmic discharge ensuing from contact between sexual bodies, when muscular tension is released in relaxation and pleasure.

Introducing an inorganic item in the circle of excitement and pleasure, for instance electronic stimulation, or accelerating the intensity and frequency of the stimuli, will result in a contraction of the psychophysical time of reaction, and a spasm of unfinished excitement will replace orgasmic discharge.

In late modern art, the idea that catharsis is the aim of art is fading, giving way to a colder, cerebral conception of the relation with art: beauty splits from pleasure, and tends to resemble a conceptual game, the space of unresolved tensions. Late modern art often resembles a frozen gesture of desensibilization.

Rather than the cathartic excitement of the modern tradition, certain artists prefer conceptual, recombinant montage. Could we say that a certain an-emotional trend marking the art-scape of the late modern age anticipated the emergence of an-affective forms of life?

In the social behavior of the generation that lives under conditions of connective precariat, aphasia seems to prevail, while verbal elaboration is compressed and accelerated to the point of provoking emotional disorders.

Dyslexia can be read as a symptom of this acceleration.

Intensified stimulation disturbs the emotional elaboration of meaning. Today, affection and sexuality seem to be wavering between loneliness and wild predatory aggressiveness, as we see in rituals of emotional detachment, virtualization, pornography, and sexual anorexia.

Sensitivity has entered a process of re-formatting. In order to be compatible with the digital machine, language must become a smooth exchange of information. The sexual imagination is now invested in the hairless surfaces of the digital image. The first digital generation is showing symptoms of emotional atrophy in the telling disconnection between language and sex. There is talk about sex in the media, in advertising, on television, and everywhere. But sex no longer includes talking, since it has become disconnected from language.

Sex is now a form of babbling, stuttering, mumbling, or screaming in a desultory way. Words are drying out.

Electronic media act as an accelerator of info-stimulation and simultaneously as a means of shrinking the sensibility of the collective psyche and of the collective skin.

Emotion: Trace and Body

What is emotion? In his writings, the psychotherapist and researcher Max Pagès tries to overcome the distinction between Freud's conception of emotion as a sign, and Reich's conception of emotion as an instinctual bodily reaction.

According to Pagès, emotion is the trace of psychological and physiological events, linked, through relevant connections, as the expressions of joy, anger, and fear.[18]

Both corporeity and culture contribute emotional values to objects, signs, and acts. A body's sensuous and sensitive reactions are linked to its cultural context. According to Pagès, there is a sensible memory through which the body records its own history of contacts, and of experiences of tenderness or violence. These are bodily traces of psychic events that are expressed more particularly as emotional inhibitions and psychosomatic issues.[19]

The skin, as it covers our body and shelters it from the external world, is also the most ancient and sensible of our sensory organs, our first tool for communication. Although it covers and encloses, the skin also opens the body to the world by bringing messages from the surrounding environment towards the mind. In the evolution of the sensory system, the sense of touch, whose organ is the skin, arises first. It is of fundamental importance, because it introduces our organism to the sensuous knowledge of the world. But the world only becomes part of our experience when our skin enters into contact with the body of the other, whether human or non-human, and warmth can flow from one organism to the other.

Society, as the space where we encounter other bodies to touch, to smell, and to see, is also the space where touching is subject to rules. Culture implies a regulation of touching and of proxemics— the way bodies are located and positioned each in relation to the other. There are cultures in which touching—starting from the relation between mother and child—is considered a commonplace occurrence. There are other societies in which touching is strongly ritualized and considered embarrassing, something to be dealt with only secretly inside the closed confines of the house, and to be reduced to a minimum in public spaces.

Noli me tangere (touch me not) seems to be the rule of behavior of modern society, where the hygienic obsession has joined and reinforced the religious one. Monotheistic religions tend to identify contact and guilt, and subject touching to rigid symbolic regulations. But in our hygienic modern times, it is a medical dissuasion, more than a religious one, that holds people back from touching and seeking pleasure. Michel Foucault described medicalization as a process that marked social spaces and functions, disciplining bodies and subjecting them to the order of economic production.

As long as touching the body was a danger for the religious soul, human beings defied the associated sense of guilt. But when the body of the other is felt as the carrier of epidemic disease, then desire itself, and not simply its expression, is repressed and reshaped. When desire becomes a danger for the body, fear begins to threaten desire from the inside, and it becomes assimilated with disease. Eroticism then takes a morbid turn, in which it is aestheticized and transferred to the realm of social taboos and individual transgressions.

Epidemics spread by contact have marked the cultural history of humanity, particularly in modern urban, crowded, and promiscuous times; think of the importance of syphilis in the nineteenth-century cultural landscape.

On the threshold of postmodernity, epidemics were greatly amplified by the mediascape, and became psycho-viral phenomena as they transferred their dangers to semiotic space, then shifted back to the sphere of emotions, and back again to the mediascape in a sort of Larsen effect of the psychosphere.

AIDS, Acquired Immune Deficiency Syndrome, is the perfect metaphor for this anthropological shift. Perhaps more than any other, this disease had massive effects on the sphere of communication. In the final decades of last century, the AIDS epidemic also spread through the media, simultaneously freezing and sanitizing the act of touching by transferring erotic energy to the realm of pornographic media, and to the endlessly postponed excitement of rituals of courting on social networks. The cultural virus has now so deeply permeated our collective psyche that we are unable to ponder its effects on the quality of our experience and in daily life. Even if the actual fear of acquiring HIV is mitigated by medical science, its cultural and psychological effects are here to stay, transformed into ritual, fashion, and lifestyle.

The Frail Psychosphere

The arts in the twentieth century favored the register of utopia in two forms: the radical utopias of both Futurism and Surrealism, and the functional utopia of the Bauhaus.

Its underlying dystopian thread, hidden in the folds of the artistic and literary imagination, borrowed from Fritz Lang's expressionism and a sort of bitter surrealism that resurfaced in the novels of Philip K. Dick.

In the second half of the twentieth century, the literary dystopias of Orwell, Burroughs and DeLillo flourished. During the transitional years from the nineteenth to the twentieth centuries, dystopia took center stage and conquered the entire field of artistic imagination. In poetry, writing, cinema, and visual art, the marks of an epidemic of mental suffering proliferated.

Throughout the late modern age, artists have been the harbingers of precariousness, which they internalized into an aesthetic of uncertainty, randomness, and excess. But in the first decade of the new century, precariousness became a social condition, pervading the labor market and workers' perception of themselves.

Precarious art is an attempt to mitigate social pain and political impotence with a kind of dystopian irony.

At the Limerick 2012 Exhibition of Visual Art, I saw *The Trainee*, a distressing work by Finnish artist Pilvi Takala produced in collaboration with Deloitte, a global network of consulting firms, and the Kiasma Museum of Contemporary Art. To realize this work, the artist spent a month as a trainee in the marketing department of Deloitte, where only a few people knew the true nature of her project. She began as a seemingly normal marketing trainee, but eventually started to engage in peculiar working methods. For example, at times, she would sit, doing nothing all day at her workstation in the consultants' open plan office space, or in the tax department library. One video shows her spending an entire day in an elevator. These acts, or rather, her absence of visible action slowly make the atmosphere around her unbearable, and force her colleagues to search for solutions, and to come up with explanations for the situation. Little by little, she becomes an object of avoidance and speculation. Her colleagues start asking her embarrassing questions, halfway between sincere interest and bewildered amusement. They address inquiries to their supervisor regarding this worker and her strange behavior. Masking laziness under apparent activity or browsing through Facebook during working hours are part of the acceptable behavioral patterns of a work community. However, sitting silently, immobile in front of an empty desk, thinking, smiling, and gazing at the wall threatens the peace of the community and disrupts the concentration of the other workers. The person who is not doing anything isn't committed to any activity, so she has the potential for anything. Since non-doing lacks a place in the general order of things, it becomes a threat to order. The degrading religion of labor is exposed here together with the uselessness of contemporary work.

In the videos *The Wind, If 6 was 9, Anne, Aki & God*, Elja-Liisa Ahtila narrates the psychopathology of relations, the inability to touch and to be touched.

In the film *Me and You and Everyone We Know*, Miranda July tells the story of a video artist who falls in love with a young man and of the difficulty of translating emotion into words, and words into touch. Language is severed from affectivity. Language and sex diverge in everyday life. Sex is talked about everywhere, but sex never speaks.

A film by Jia Zhang-Ke, entitled *Still Life* (*Sanxia Haoren*) and produced in Hong Kong in 2006, shows the progressive devastation of contemporary China. The predominant color is a rotten, grayish, violet green.

The story is simple, but cruel. Huo Sanming returns to his native village in the hope of finding his wife and daughter, whom he had left years earlier to go and look for work in a distant northern mine. His village along the riverbank of the Yangtze no longer exists since the construction of the Three Gorges Dam. Many villages have already been erased from the map, houses, people and streets covered in water. Yet the building of the dam proceeds, the destruction of villages continues, and the water keeps rising. Huo Sanming arrives in this scenario of devastation and rising water, and he does not find his wife and daughter. So his search begins. He looks for them while groups of workers armed with picks tear down walls, explosives demolish buildings, and the landscape is transformed into a huge sprawl of garbage.

After a long quest Huo Sanming finally finds his wife. She has aged and has been sold by her brother to another man. They meet in the rooms of a building as it is being demolished and, in whispers, they talk about their daughter, their heads bent. A dark green spaceship hovers in the background where bricks and iron are spattered onto a shit-colored sky. In the last scene of *Still Life*, a tightrope walker inches along a rope from the roofs of a house towards nothingness, against a background that recalls the dark surrealism of Dali's bitter canvases. *Still Life* is a lyrical account of

Chinese capitalism from the inside out, from the standpoint of submerged life.

The Corrections, a novel by Jonathan Franzen, tells of humanity's minute psychopathological shifts and psychopharmacological adaptations as it is increasingly devastated by depression and anxiety. The novel shows people attempting to adjust to an existence that appears and pretends to be normal, while their brains are unable to deal with the chaos that surrounds them and that governs their intimate lives. The title of the novel refers to the adjustments needed in a volatile stock market in order to avoid losing the money invested in private pension funds that might suddenly disappear.

Franzen narrates the trials of an aging couple in the Midwest who have gone crazy as a result of decades of excessive labor and conformism. The term corrections, here, also refers to the slow, unstoppable slide towards sexual turn-off, and to the horror of old age in a civilization of competition, or the horror of sexuality in the world of puritanical efficiency.

In this 2001 novel, Franzen digs deeply into the folds of the American psyche to describe in minute detail the reduction of the American brain to a pulp, showing the depression and dementia that result from prolonged exposure to the psychic bombardment of stress from work, apathy, paranoia, puritan hypocrisy, and the pharmaceutical industry that encircles all of it. Franzen shows the psychic unmaking of those who are caught in the claustrophobic and illusory shell of economic hyper-protectionism, and reveals the infantilism of those who pretend to believe, or perhaps truly believe in the fulsome Christmas fairy tale of compassionate liberal cruelty. By the end of a long awaited Christmas dinner, as the psychopathic family happily gathers together, the father tries to commit suicide by shooting himself in the mouth. He does not succeed.

Yakizakana No Uta starts with a fish in cellophane wrapping on a supermarket shelf. A boy grabs it, and takes it to the register. He pays, leaves, puts it in his bicycle basket, and cycles home.

"Good morning, Mr. Student. I'm very happy to be with you. Do not worry, I'm not a fish who complains," says the fish as the student pedals briskly. "It's nice to make the acquaintance of a human being. You are extraordinary beings. You are almost the masters of the universe. Unfortunately you are not always peaceful. I would like to live in a peaceful world where everyone loves one another and even fish and humans shake hands. Oh, it's so nice to see the sunset. I like it ever so much." Saying this, the fish becomes emotional and flops around in the cellophane bag inside the basket. "I can hear the sound of a stream … I love the sound of streams, it reminds me of my childhood."

When they get home the boy unpacks the fish, puts it on a plate, and throws a little salt on it. As the fish grows excited, and proclaims "Ah! I like salt very much, it reminds me of something," the boy places it on the oven grill and turns the knob.

The fish keeps chatting, "Oh, Mr. Student, it's nice here, I can see a light down there … I feel hot … hot," until its voice becomes hesitant.

Then it starts singing a song, increasingly feebly and unconnected, reminiscent of Hal in *2001: A Space Odyssey* as his wires are unplugged.

Although Sakamoto's *Yakizakana No Uta* was perhaps the most harrowing animation film I saw at the CaixaForum Barcelona during the *Historias Animadas* festival in June 2006, I perceived a common tone of ironic cynicism and despair running through all of the works presented at the festival.

Ruth Gomes' *Animales de compania* uses ferocious images to tell the story of a generation of well-dressed anthropophagi, young beasts in ties. The latter are endlessly fleeing to avoid being caught by fellow colleagues, friends, and lovers who wound, kill, and eat them as soon as they fall into their grip, with terrorized smiles and dilated eyes.

The art I am describing is not an art of denunciation. The terms *denunciation* and *engagement* are no longer meaningful when you are a fish reaching the point of being cooked.

Even though they keep using expressions taken from last century's lexicon, artists of the twenty-first century no longer show that kind of energy, perhaps because they are scared by their own truth. Artists no longer search for rupture. Instead, they seek a path that may lead to equilibrium between irony and cynicism, a way to suspend the execution, at least for a moment.

Geo-Cultures: Skin and Imagination

This book traces the mutation that the globalization of technology and media has produced in the living body of cultures. It also describes the anthropological mutation that follows the wide diffusion of electronic technology and the digitalization of social communication as it affects different cultures, with their own internal conflicts.

Contrary to the presuppositions of the theory of the clash of civilizations, in my opinion, identities do not exist as such. They are fragile and changing constructions, based on the social history of human groups, and on the effects of flows of techno-psycho stimulation.

Civilizations are not homogeneous blocks of consistent identities, but rather the space of a continuous negotiation between differences.

If we look at the geopolitical and geo-cultural map of the world, we see that most current conflicts are occurring within so-called civilization. In his book *The Clash of Civilizations* (1996), Samuel Huntington writes that the Christian West is destined to clash with Islam—but the experience of the last ten years has shown that the main source of war is the opposition between Sunni and Shia Islam.

Far from being the natural evolution of long-term legacies, identities are shifting, fragmented, and continuously reshaped by the flows of media that cross through them.

Strictly speaking, civilizations do not exist. The transversal factors of mutation, that is, technology, the Internet, financial collapse, rather than pitting civilizations one against the other, are permeating different cultural landscapes with similar colors.

The word *civilization* does not define a political homogeneity. The transition at hand does not oppose races or religions, the usual markers of civilization. Instead, it sets technologies and economic lifestyles in opposition within a common framework marked by exploitation, social misery, and mental suffering.

This book intends to trace the molecular evolution of geo-cultures, and to do so, will focus on identities simply as temporary constructions.

In order to draw a map of this mutation, I will follow two different pathways. I will begin by tracing a phenomenology of the skin, and continue with a genealogy of global imagination.

This phenomenology of the skin draws a map of the multifarious approaches to perceiving the body of the other, where *culture* is reframed as the sphere of social imagination, that is, both as perception and projection of the environment.

The following chapters will focus on sensibility and social imagination in the modern age.

2

Global Skin: A Trans-Identitarian Patchwork

An uncanny geography, uncanny in Freud's sense of "that class of frightening which leads back to what is known of old and long familiar." [… The] "uncanny" is actually the *unheimlich*, the unhomed or that which is not at home. Both its frightening and its familiar qualities come from its awkward relation to being not at home, to the strangeness which that condition assumes.

— Irit Rogoff, *Terra Infirma: Geography's Visual Culture*

Skin is basically a two-layered membrane. The lower, thick spongy dermis, one to two millimeters thick, is primarily connective tissue, rich in the protein collagen; it protects and cushions the body and houses hair follicles, nerve endings and sweat glands, blood and lymph vessels. The upper layer, the epidermis, is 0.07 to 0.12 millimeter thick. It is primarily composed of squamous, or scalelike, epithelial cells, which begin their lives round and plump at the boundary of the dermis and over a 15-to-30 day period are pushed upward, toward the surface, by new cells produced below. As they rise, they become flattened, platelike, lifeless ghosts, full of protein called keratin, and finally they reach the surface, where they are ingloriously sloughed off into oblivion.[1]

Our skin is interposed between the world, and ourselves and acts as a sensitive processor of worldly experience. It is continuously re-generated, emerging at the surface, aging, decaying and finally disappearing as it melts into air, forgotten. But the sensitive data it has recorded does not die or disappear: it is *stored* in the brain, transformed into memories, and turned into sensitive expectations. The skin feeds the brain with perceptions of the world, but conversely the brain supplies the skin with sensitivity, aesthetic inclinations, and tendencies: desire. Desire is not the need for something, but the sensible creation of the world as an aesthetically meaningful environment.

Cultural history, therefore, has something to do with the history of the skin. Experiences deposit sediments in our bodily memory—these sediments in turn give shape to our selves.

However, as Max Pagès suggests in *Traces ou sens*, we should not infer that there exists a collective memory or a collective skin that would define an ethnic or cultural identity. Bodily memory is singular, not collective, and the interplay between individual selves is not pre-determined by any super-individual identity. Collective identity—an ethnos, a *volk*, a nation, and so on—is nothing but a fiction, an arbitrary fixation of identification processes.

The current transition from a conjunctive to a connective environment is reshaping the perception of self in a context defined by different forms of life.

In the English language, there is a slight but very profound semantic difference between the words *sensibility* and *sensitivity*. Sensibility refers to the ability to detect meaning, that is, morally and conceptually meaningful implications in non-verbal enunciations such as gestures, hints, and existential situations. Sensitivity, on the other hand, refers to the ability to detect meaningful implications in tactile perceptions, in epidermic stimuli, and in sexual innuendo. These two words provide access to the spheres of aesthetics and eroticism.

The mutation we are experiencing today is provoking a painful dissonance between the spheres of sensibility and sensitivity, a dissonance that is felt in many ways. As the conscious organism sensuously opens itself to other bodies, this connection resonates with the pleasure raised by artificial, aesthetic signs. Eros and art interfere, creating disturbances, overlapping, and contamination.

Natural and artificial dimensions cannot be dissociated from the point of view of emotional response. Erotic languages obviously imply aesthetics, and clearly, aesthetic production stimulates erotic effects in the social sphere.

The Inner Touch

In his book *The Inner Touch*, Daniel Heller-Roazen traces the philosophical sources of the perception of self as sensitive self-awareness,

> the experience of the one sense shared by all the individual senses and felt, however faintly and however intermittently, in all sensation: the sense of sensing [...].[2]

Diane Ackermann, in *A Natural History of the Senses*, writes:

> What is a sense of one's self? To a large extent, it has to do with touch, with how we feel. Our proprioceptors (from Latin for "one's own" receptors) keep us informed about where we are in space, if our stomachs are busy ...[3]

Kant speaks of apperception to define the self-perceived unity of consciousness, the space in which cognitive activity can take place, and Heidegger speaks of *Da-sein* to define the perceivable presence of the existence of the self.

Heller-Roazen reframes the problem of the unity of consciousness through *aisthesis*, the sense of sensing.

> The distant origin of the modern "synesthesia," the Greek term was no neologism when the thinkers of late Antiquity bestowed upon it a technical sense in the doctrine of the soul. [...] Formed by the addition of the prefix "with" (*sun-*) to the verb "to sense" or "to perceive" (*aisthanesthai*), the expression in all likelihood designated a "feeling in common," a perception shared by more than one. [...] At this point in the development of the Greek language, the term applied to the communal life of many, and its meaning lay far from the one that would later be attributed to it by the commentators.
>
> One of the earliest indications of a shift in the sense of the expression can be found in the medical literature that flourished after the beginning of the Christian era. It has been noted that Galen, for instance, employs *sunaisthēsis* to designate a sensation in common not in that it is shared by many but in that it reaches a single body all at once, while consisting, in effect, of multiple physiological affections.[4]

The shift from the notion of *synaesthesis* as a perception shared by many, to the notion of synaesthesis as inner touch highlights the relation between the sensitive consciousness of the self and the proxemic consciousness of others. Self-perception does not only imply the synaesthetic act of the inner touch, the singular sensing of the body of the self, but also implies the others who are perceiving your body and your existence in their own space. Self-consciousness and proxemics mutually interfere, and the processes of individuation and of identification are narrowly linked.

According to the anthropologist Edward Hall, proxemics is a discipline that studies non-verbal communication and the disposition of bodies in space.

Today's connective transition is transforming the conditions of social proxemics, i.e. the arrangement of bodies in space. As the media transform the way people interact in space, the disposition of bodies in the urban landscape shifts.

Proxemics also defines the physical substratum of what Gilbert Simondon calls *individuation*. By that word, Simondon refers to the separation of the individual from the undifferentiated continuum of the environment, as well as to the specification of the self as a conscious and sensible organism. The individual, whom the modern philosophical tradition has generally considered the premise of every discourse about society, should instead be rethought as the product of social relations, and as the result of a process of differentiation from the pre-individual reality of organic matter.

Liberal thought considers social relations to be the interaction between individuals whose existence is taken for granted. On the contrary, individuals are instead better understood as the product of a maze of relations and implications. It is only against the backdrop of this maze that the self-perception of the individual emerges. Sensitivity, the ability to feel the meaning of the other, is what leads to the differentiated perception of the self.

The following pages will map out some of the ways processes of identification function in different psycho-cultural contexts. I do not pretend to any kind of exhaustive taxonomic description, neither am I attempting to draw a detailed psycho-cultural carto-graphy. Instead, by simply meandering through the contemporary culture-scape, I am trying to shed light on some crucial passages in the history of sensibility to prepare the ground for a deeper under-standing of the current technocultural mutation.

Psycho-cultures are located in a geographical dimension, although the geographical location has very little, if anything, to do with belonging, as Irit Rogoff argues in *Terra Infirma*:

> My inquiry does not attempt to answer the question of a location
> for belonging; it is by no means prescriptive since I have no idea
> where anyone belongs, least of all myself. It is, however, an
> attempt to take issue with the very question of belonging, with its
> naturalization as a set of political realities, epistemic structures
> and signifying systems.[5]

I will begin my geo-psycho-cultural wanderings with an Indian writer who has particularly, and almost obsessively focused on the problems of identity. The contemporary, post-colonial globalization of culture seems to me a process of de-sensibilization that is con-current with the rise of a new process of sensibilization. I will then return to the sources of modern sensitivity. Starting from an inves-tigation of the Christian concepts of love and eroticism, I will arrive at an analysis of romantic and late romantic aesthetics. This will allow me to assess the sex appeal of electronics, and the digital effect that the contemporary process of de-sensibilization is having on contemporary psycho-culture. Finally I will describe this process of desensibilization, or the mutation of sensibility, as it relates to daily life in countries such as Japan and South Korea where digitalization has deeply affected the psychosphere.

Homelessness and Identity

Although I will be speaking of identities in the following pages, my intention is not at all to celebrate them, or to accept their definition as a natural given. I'm interested instead in deconstructing their genesis, and critiquing their use for political and psychological ends. For the word *identity* generally fixes a stage in the process of identification, and hypostatizes it into a natural given.

The expression *cultural psychology* is borrowed from the writings of Sudhir Kakar, the Indian psychoanalyst who has written extensively about Hindu sexuality and mythology. Cultural psychology is the discipline that retraces processes of identification in different cultural environments, avoiding fixing and hypostatizing particular stages of identification.

The literary and theoretical writings of V. S. Naipaul are the perfect starting point for a reflection on the traps inherent in the concept of identity, as his works trace the emotional loneliness of post-colonial nomads as they move from individuation to identification.

Cultural nomadism has become a widespread condition in an age of globalized labor markets and media outlets. Nomadism, the process of cultural deterritorialization and of its subsequent psychological uprooting, deeply affects the perception of one's self, causing psychological suffering while simultaneously opening new perspectives of imagination and identification.

Naipaul's literary experience provides for an incisive overview of our contemporary cultural mutation.

Born to a family of Southern Indian origin in Trinidad, the Caribbean island where many Indian indentured workers had migrated during the first part of the twentieth century, Naipaul was raised in the Hindu tradition, but went to London in his youth to study at Oxford, eventually becoming one of the most important English writers of our time. His books narrate his travels, his encounters with people from very distant cultural environments, and describe various cultural contexts—the Indian, the Islamic, the European, the South American, and the African—from a deterritorialized point of view, that is, one external to the multifarious situations in which he happens to be involved.

Through his writing, Naipaul attempts what he calls "a synthesis of the worlds and cultures that had made me."[6]

Fears, inhibitions, shyness, and aversions emerge from his own psycho-cultural background to mark his judgments, his gestures, and the specific irony and sarcasm that he inserts between himself and others. The avoidance of touch, so deeply rooted in the Brahminical lifestyle and character, is constantly present in his words as they denounce the embarrassment provoked by the body, the food, the beliefs, and the sexuality of the other.

His books express a kind of repulsion aroused and instilled in him by human physical presence.

Paradoxically, Naipaul might also be viewed as a witness to global eros, although one could say that his writing is the exact contrary of eroticism, focusing on the erotic (ill)feeling and (dis)pleasure provoked by the accessibility of other bodies.

From his books, it seems to me that although Naipaul deeply despises the individuals and human groups that he meets as he travels, the person that he seems essentially to despise is himself: his own skin, his own body, his own face, his own story, and his own past.

Naipaul even hates language, the English language, which he masters in a perfect and crystalline way but which does not belong to him. He hates his own linguistic matter, although he writes bowing down and crawling into the shining heights and obscure intricacies of the colonizer's tongue, which he has received as a poisonous gift and transformed into something that is absolutely intimate.

> The world is illusion, the Hindus say. We talk of despair, but true despair lies too deep for formulation. It was only now, as my experience of India defined itself more properly against my own homelessness, that I saw how close in the past year I had been to the total Indian negation, how much it had become the basis of thought and feeling.[7]

Homelessness is the key word for Naipaul's experience and work. It is possibly the key word for our time, a time for which Naipaul is a privileged decoder. Homelessness is the condition of those who do not have a dwelling, a homeland, or an identity, and who also have no place to which they can return.

The psychological and political universe of the post-colonial age is filled with, and radically disturbed by the lack of belonging, the desire to belong, as well as the emptiness and deception of belonging. The root of this anguish can be located in the space of sensitivity and sexual proxemics. It is linked to the vanishing of the body in the sphere of virtual globalization, and to the obsessional return of the body as unfulfilled desire.

> To us, without a mythology, all literatures were foreign. Trinidad was small, remote and unimportant, and we knew we could not hope to read in books of the life we saw about us. Books came from afar; they could offer only fantasy. I went to books for fantasy; at the same time I required reality.[8]

In the colonial world, the center was clearly the colonizing West, while the colonies—both Trinidad and India—were relegated to the role of periphery. Post-colonial reality has jeopardized the centered perception of the world, and today, everyone is left with the question: "Where is here?" There is no possible answer to that question, since we've lost the reference that previously existed, and identity is at last exposed as deceiving, an infinite search for something that does not exist.

> The customs of my childhood were sometimes mysterious. I didn't know it at the time, but the smooth pebbles in the shrine in my grandmother's house, pebbles brought by my grandfather all the way from India with his other household goods, were phallic emblems: the pebbles, of stone, standing for the more blatant stone columns. And why was it necessary for a male hand to hold the knife with which a pumpkin was cut open? It seemed to me at one time—because of the appearance of a pumpkin halved downward—that there was some sexual element in the rite. The truth is more frightening, as I learned only recently [...]. The pumpkin, in Bengal and adjoining areas, is a vegetable substitute for a living sacrifice: the male hand was therefore necessary. In India I know I am a stranger; but increasingly I understand that my Indian memories, the memories of that India which lived on into my childhood in Trinidad, are like trapdoors into a bottomless past.[9]

It is very interesting to define belonging and identity as "trapdoors into a bottomless past." In the sphere of contemporary globalization, this trapdoor of identity has lost its meaning and its foundations. Paradoxically, however, it is the very uprooting and loss of living memory that feed a nostalgic desire for an identity that never really existed in the first place. The historical exposure of the groundlessness of identitarian roots has in fact excited the need to belong.

> I began to feel when I was quite young that there was an incompleteness, an emptiness, about the place, and that the real world existed somewhere else.[10]

Of course, the real world is always somewhere else. But such outlandishness is difficult to assimilate consciously, and even harder to accept. That is why we tend to identify the real world with the environment that surrounds us. But identity is reframed in the cultural transition currently underway. Deterritorialization, the virtualization of social space, and the replacement of physical experience with simulation have produced a new dimension of synthetic identity for which America is the name.

Becoming Americans

The Nazis' conception of eugenics was based on exclusion. Only those who belonged to the Aryan race were allowed a privileged access to the world that was projected for the future, a world cleansed of all impure residuals. Jews, Romani, Africans, homosexuals, and communists were all barred from the Nazi paradise. American identity, on the contrary, is moved by an inclusive intention. As soon as its land was colonized, everyone could become an American, and every ethnical and linguistic group was admitted into the American ecumenical world, into the melting pot where different identities blended and forged the American one.

Simultaneously, however, the cultural identity of American history is based on the puritanical cult of the Word of God—as

Samuel Huntington has argued in his 2004 book *Who Are We?: The Challenges to America's National Identity.*

These two things are not contradictory. Puritanism is not an identity, but a process of cancellation of past identities. This was already the case in the seventeenth century, when English Puritans left the British islands and the European continent in order to forget about their past as part of a church that had descended from the impure Roman Church. What the Pilgrim Fathers did first was to abandon, forget and erase their link to the impurity of the past. Only then did they found a new homeland, which was not a continuation of past history, but an empty space where the Word of God could create institutions and forms of life. This opened a neo-human dimension in North American space.

Inclusion in the neo-human results from a multilayered re-formatting of those individuals wishing to be admitted to the new world.

It implies a cultural, linguistic, and emotional process of re-formatting that interweaves with a technological re-formatting to enable the functional integration of individuals into a connective universe.

This integration does not take the form of cultural, linguistic, and emotional assimilation. Rather, it implies that individuals become operationally compatible with the connective rationale. Expectations and desires are reshaped through this process of operational compliance.

While the exclusionary quality of ethnic nationalism is based on an inaccessible identity, American identity resides precisely in its expansive becoming. American history is an experiment in a synthetic construction that progressively absorbs organic identities, consequently marginalizing or cancelling what cannot be bent to its operational principle, or what cannot be translated into its digital language.

American exceptionalism consists, firstly, in the fact that the country inherited the essential legacy of all the cultures of the world, while simultaneously evacuating their sensory, historical materiality, and the weight of their lived experience. Secondly, it consists in the claim to economic, political, and military self-reliance that defines the strategy and vision of America's leading class. By America, here, we are not referring to the North American territory subject to the jurisdiction of the United States of America, but to a pervasive anthropological principle that defines the way of life and self-perception of the global class connected through the circuit of the global economy.

The entire population of the planet since the Second World War has, in fact, undertaken the process of becoming American, thanks to its mostly willful adhesion to the mythology propagated by Hollywood, the global advertising industry, and the political machinery of human rights. To become American means, first of all, to be part of the global circulation of goods and images. Secondly, it implies the possibility of freeing oneself from the heaviness of tradition, belonging, subjection, and tribal rules of women's oppression. But it simultaneously implies losing contact with the concrete experience of affective singularity, and purifying the social existence of anything that might obstruct a perfect integration into the productive cycle.

There is a plentiful literature about the double meaning of becoming American. On the one hand, there is the euphoric feeling of freedom that everyone can experience walking in the streets of a city where nobody looks at you or bothers you, the euphoric perception of a sort of emptiness that opens the door to adventure, self-discovery, initiative, and success. On the other hand, there is the sense of loneliness, and the impoverishment of shared sensibility. Innumerable people have experienced this in the age of modernity, and many writers, particularly women writers, have narrated this process of becoming American. Three such writers, Barati Mukherjee, Chitra Banerjee Divakaruni, and Jhumpa Lahiri, have compared the thickness of their Indian memory with the experience of their American present, and its accompanying difficulties.

The Indian imaginary they carry in their memory is packed with myth, ritual, forms of life, affection, and emotion, and through the point of view of gender they each differently retrace the transformation of their emotional landscape, the shifts in their relation to food and to the city, the alteration of how bodies are arranged in space, their interactions, looks and gazes, as well as the entire social proxemics.

In *The Interpreter of Maladies*, Jhumpa Lahiri gracefully embroiders upon the painful yet exciting encounter between her heavy Indian existential universe, and the anxious lightness of the American universe she is joining. The book is a collection of short stories. One of them, *Mrs. Sen's*, narrates the daily life of a woman who moves from Calcutta to Boston in order to live with her husband, a professor at a university. Her days are empty, suspended in an atmosphere of loneliness and meaningfulness. She compares the affective density of the world she has left behind her, the extended family, the crowded city, and the childhood memories with the emotional desert of her days in the new world, where work is the only common ground of exchange and of understanding.

The short story that gives its title to the book, *The Interpreter of Maladies*, reflects upon the trauma of self-transformation as it occurs through immigration and the series of broken identities that result, a collection of multiple anchorages. For loss of identity can, at times, lead to the discovery of a wide range of previously unimaginable possibilities.

Lahiri's stories show how members of the diaspora struggle to maintain their traditional culture as they invent new lives for themselves in foreign cultures, hybrid realizations that are simultaneously exciting and frightening.

This becoming American can be viewed as a passage to the sphere of the neo-human. It involves a process of purification from the residuals of identity and belonging that is not exactly the erasure of memory, but instead a kind of re-coding and re-semiotizing of its contents.

Arjun Apadurai, in his 1996 book *Modernity At Large: Cultural Dimensions of Globalization*, speaks of *diasporic public spheres* to refer to the shift from one imagination to another, and to the overlapping of different layers of memory. But this co-existence and overlapping of different cultural worlds, which Apadurai views as a pulverized and expanded sphere of modernity, is re-coded and re-functionalized by a sort of abstract machine of cognitive operationality.

Only those who can comply with the digital mind that is embedded and objectified in the techno-economic universe can be fully integrated in the neo-human sphere. This modernity at large, which is incorporated into techno-economic compatibility, infiltrates every fragment of the life of those freed from traditional and colonial constraints. It shapes and concatenates individual cognitive response, preserving cultural differences to some extent.

Sin and Pleasure

The Christian myth of incarnation implied a radical break with the Jewish legacy of an inaccessible God whose name cannot be spelled, and whose image cannot be portrayed. In the Christian tradition, since God became human, implications regarding flesh in the religious sphere gave a new meaning to the concept of love.

The opposition of the finitude of the flesh and the infinite of the relation with God is one of the basic obsessions of Western culture. It forms the precondition for the kind of *romantic hysteria* that emerged in the Christian sphere during modern times.

The Augustinian opposition between *eros* and *agape* involved a depreciation of desire as self-centered and egoistic. Christian culture, which was originally based on love and compassion, came to be transformed into a culture of renunciation. Eros was stigmatized as the selfish search for lust, while the altruistic side of pleasure was forgotten and denied, paving the way for a phobic conception of the individual. The dissociation of love and eroticism reduced compassion to a merely moral sentiment, whereas the etymological root of the word means *shared perception.*

The separation of pleasure and interest, which can be viewed as a pathology of self-love, was the condition for the separation and opposition of interest and pleasure, which prepared the cult of accumulation and capitalist valorization in modernity.

Notwithstanding the infusion of erotic pleasure with guilt, which emerged in the Christian sphere after Augustine, at the beginning of the second millennium a new sensibility based on the notion of love as spiritual knowledge rose from the meeting of Christian and Muslim cultures in the western Mediterranean area of Andalucia. Here we find the origins of courtly poetry (or *poesia cortese*, according to the Tuscan denomination of the *Dolce Stil Novo*) that flourished from the twelfth to the thirteenth century along the Mediterranean coast from Seville to Florence.

During those centuries, the meeting of two religious cultures that were in themselves rather sexually phobic and intolerant led to the emergence of a new form of imagination and expression, paving the way to the modern conceptions of freedom and sensuousness.

In the works of the philosopher Ibn Arabi from the turn of the twelfth century, eros and agape, far from being opposed, mutually feed and exalt each other. In the same period, the poet Ibn Hazm, author of *Tawq al-hamamah* (*The Ring of the Dove*), composes poems that praise erotic love as a condition of spiritual elevation.

Love mobilizes energies, this is well known. And these energies can provoke destabilizing effects. This is why monotheistic religions, which are linked to patriarchal power and tend to protect its stability, are afraid of women's sexuality

According to Fatima Mernissi, in the Muslim cultural sphere "women are feared because uncontrolled sexuality is regarded as destabilizing to the community—and female sexuality is regarded as the most dangerous. The Arabic word *fitna* means disorder or chaos, but it can also refer to a beautiful woman, thus demonstrating a link between women and instability."[11]

But, as Ibn Hazm writes, desire can be seen as a force of elevation and of spiritual mobilization.

> And I suppose desire
> Is like a coal,
> That feeds upon the fire
> Still in my soul.[12]

Retracing the origins of passionate love as a defining feature of modern Western sensibility, Denis de Rougemont argues that romantic love as we know it is a relatively recent development, one peculiar to Western Culture, and that usually entails an illicit rapport such as adultery as well as an inextricable psychological link to death.

Commenting on the Norman poet Béroul's adaptation of the myth of Tristan and Isolde, de Rougemont writes that:

Passion means suffering, something undergone, the mastery of fate over a free and responsible person. To love love more than the object of love, to love passion for its own sake has been to love to suffer and to court suffering all the way from Augustine's *amabam amare* down to modern romanticism. Passionate love, the longing for what sears us and annihilates us in triumph—there is the secret which Europe has never allowed to be given away; a secret it has always repressed—and preserved![13]

And also:

No passion is conceivable or in fact declared in a world where everything is permitted. For passion always presupposes subject and object, a third party constituting an obstacle to their embrace—a King Mark separating Tristan from Isolde—the obstacle being social (moral, conventional, even political) to such a degree that we even find it identified, at its limit, with society itself, though it is generally represented by a dramatis persona, in accord with the requirements of narrative, the rhetoric of romance.[14]

According to de Rougemont, a double meaning is involved in the *topos* of passionate love, the ambivalence of pleasure and sin.

Since the second century of the second millennium, in some areas of the Christian world, namely Andalucia, Languedoc, and Tuscany, love has emancipated itself from the passionate chains of sin, and is deployed as a path to elevation.

In the words of Guido Guinizelli, who is considered the initiator of the Italian school of *Dolce Stil Novo*, love is free from the schizo-phrenic duplicity of sin and pleasure and is conceived as a channel to divine enlightenment. Looking at the beauty of another body is a way to get closer to the vision of God. It implies the erasure of the opposition between the earthly and the spiritual, and the end of the traumatic separation between eros and agape.

> Within the gentle heart Love shelters him,
> As birds within the green shade of the grove.
> Before the gentle heart, in nature's scheme,
> Love was not, nor the gentle heart ere Love.
> For with the sun, at once,
> So sprang the light immediately; nor was
> Its birth before the sun's.
> And Love hath his effect in gentleness
> Of very self; even as
> Within the middle fire the heat's excess. [...]

> God, in the understanding of high Heaven,
> Burns more than in our sight the living sun:
> There to behold His Face unveiled is given;
> And Heaven, whose will is homage paid to One,
> Fulfills the things which live
> In God, from the beginning excellent.
> So should my lady give
> That truth which in her eyes is glorified,
> On which her heart is bent,
> To me whose service waiteth at her side.

> My lady, God shall ask, "What daredst thou?"
> (When my soul stands with all her acts review'd;)
> "Thou passedst Heaven, into My sight, as now,
> To make Me of vain love similitude.

To Me doth praise belong,
And to the Queen of all the realm of grace
Who slayeth fraud and wrong."
Then may I plead: "As though from Thee he came,
Love wore an angel's face:
Lord, if I loved her, count it not my shame."[15]

Facing God, Guido Guinizelli does not beg forgiveness for his love of a woman. He believes that he committed no sin, because the woman looked like one of God's angels.

Christian love is taken in a sort of double bind, insofar as it is bound by contradictory injunctions such as "Love your neighbor as yourself," and "Thou shalt not commit adultery." Free love conflicts with the institutionalized bond of marriage.

In Beroul's version of the legend of Tristan and Isolde, Tristan—deceitful in spite of himself—betrays Mark, and suffers for his own betrayal. His passion debases rather than elevates.

Christianity's simultaneous exaltation and condemnation of sensuousness encourages the desire for knowledge, and is therefore a gateway to the humanist revolution.

Passion is that form of love which refuses the immediate, avoids dealing with what is near, and if necessary invents distance in order to realize and exalt itself more completely.[16]

Here we find the symptom of a self-feeding malady and the origin of a feverish discourse that will result in the romantic impulse for the continuous overcoming of limits, as well as the psychological motivation for the unceasing investment of energy, and for accumulation and progress.

The extreme form of this paradox is to be found in Catholic mysticism. Pleasure and torture are mutually intertwined in ecstatic vision, which makes the cross the universal aesthetic paradigm. Think of Saint John of the Cross, the extreme lover whose flesh is consumed in a fever of mystical desire. Think of Zurbarán, Murillo, and particularly El Greco, the painters who established the model of the Sacred Heart of Jesus. In this popular mythologem of modern Mediterranean religiosity, the physical representation of the tearing asunder of Jesus' flesh echoes the sadistic mixture of piety and pleasure that people experienced when attending the public torment inflicted on the victims of religious spectacle.

The Christian process of modernization simultaneously implied the repression of the wild side of the sensorium—touch, smell, sexual pleasure—and the regulation of the civilized senses— hearing and vision. The visualization of eroticism was the prevailing trait of the modern regulation of sensuousness.

The Flesh Is Sad

Romantic aesthetics reconfigured the very idea of beauty according to the ambiguous schema of the notion of the sublime. This reconfiguration ran parallel to the modern reframing of erotic sensibility.

Before the romantic age, in the late eighteenth century, the geometric spirit of rationalism (*esprit de géometrie*) coincided with the geometric spirit of libertinism. In the wake of the Restoration, a new sense of historical tragedy and human weakness obfuscated the scene of the Enlightenment, and a feeling of exhaustion and of in-distinction invaded sensibility. For Schelling, the thinker who most fully expressed the twilight of the enlightened spirit and the emergence of a more nuanced and blurred sensibility, the sublime was a way to express a rising awareness of infinity.

When confronted with the cosmic infinite, our senses are induced into a condition of confusion and panic. In Greek mythology, Pan is the divinity who symbolizes the infinity of nature. As soon as the geometric rationalism of the Enlightenment receded, leaving room for the disturbing experience of boundlessness, Pan was back.

Sublime and panic are bordering concepts. Panic is the opening of consciousness before the infinity of nature and the failure of rational filters of experience, and the sublime is the epiphany of the unknowable.

Pan submerged perception almost to the point of drowning; his sudden arrival provoked fear and pleasure, malaise and excitement. This panicked perception appeared clearly in the paintings of Caspar David Friedrich, or in the poems of Percy Bysshe Shelley.

In the late modern language of psychopathology, doctors use the word *panic* in order to define a new sort of pathology whose symptoms are an accelerated heartbeat, intense perspiration, shortness of breath, confusion of mind, anxiety, and shaking.

In a sense, the shift from the rationalist *esprit de géometrie* to the romantic return to the *esprit de finesse* can be seen as a modulation of the dilemma opposing Leibniz and Spinoza, the founding fathers of modern philosophy: recombinant rationale of the finite divisibility of matter (the monad as information), versus the infinite substance of the universe as inexhaustible source of experience.

I'm referring here to the genealogy of the main subjects of this book, that is, the dilemma between conjunction and connection, the historical shift from the conjunctive mode to the connective mode of social communication, and the epistemic approach to experience. We can re-read the history of modern culture—philosophy, and aesthetic sensibility—from the point of view of this dilemma, which can only now, in the age of digitalization, be fully appreciated.

From where does the romantic passion for the sublime come? Rather than the sublime, beauty was the value that classical art cherished. Classical art—and the humanist return to classicism—identified beauty with splendor and light, with symmetry and proportionality, with the worldly finiteness and measurability of the body and of the objects created by human artifice.

Monotheistic religions devaluated sensuousness as the region of sin and temptation, so beauty became an ambiguous concept, since beauty diverted man's energies from the true object of spiritual life, the infinity of transcendence.

The devaluation of the senses is inherent to monotheistic religious cultures. Whereas in the sphere of pagan cultures, the *entheogenic* experience of drugs, hallucinated visions, and mysticism are part of religious sensibility, in the sphere of monotheist religion, mysticism is suspicious because it mixes the sensuous experience of pleasure, exhilaration, and excitement with the spiritual knowledge of God.

The erotic perception of beauty, in this context, is demonized, veiled, forbidden, since it is identified with a sinful distancing from God.

After the Protestant schism, something happened in the Christian cultural domain. While the Protestant Reformation established an aesthetic space of severity and essentiality, the Catholic baroque entered the aesthetic space of phantasmagoria, of the wonderful, and of the spectacular product of human artifice, charged with exciting the mind and leading the imagination towards the vision of God.

But baroque aesthetics were marginalized by the modern culture of the northern bourgeoisie. It was there, in the North, far from baroque phantasmagoria, that the romantic sensibility thrived, that the sublime met the horrid, and that excitement mixed with agony. Eventually, romanticism melted away and passed into the sphere of symbolism.

> Oh Beauty! Do you visit from the sky
> Or the abyss? Infernal and divine,
> Your gaze bestows both kindnesses and crime [...]

The perception of aesthetic beauty ambiguously and intimately blended with the distressing perception of erotic sensuousness.

> The panting lover bending to his love
> Looks like a dying man who strokes his tomb.

> What difference, then, from heaven or from hell,
> O beauty, monstrous in simplicity?
> If eye, smile, step can open me the way
> To find unknown, sublime infinity?[17]

Symbolism emerged from the sublimation and exhaustion of the flesh. Mallarmé's azure appeared from a frigid mental ecstasy that took the place of sensuous pleasure.

> My soul, calm sister, ascends toward your brow
> Where an autumn that's scattered with russet dreams now,
> And toward your angelic eye's wandering heaven
> Ascends, as in a melancholy garden
> A white jet of water faithfully sighs
> Toward October's pure, pale, and compassionate skies
> That mirror in pools their infinite languor

And, on the dead water where anguished leaves wander

Driven by wind, furrowing a hollow,

Let the sun be drawn out in a long ray of yellow.[18]

The autumnal frigidity of extinguished sensuousness permeated symbolist poetry.

The flesh is sad, alas, and there's nothing but words!

To take flight, far off! I sense that somewhere the birds

Are drunk to be amid strange spray and skies.[19]

The flesh is sad because the body is exhausted by its flight towards the infinity of experience. In the symbolist sphere, sweetness, pleasure and sensuousness were completely absorbed by language, by the sounds of words, and by the trans-mental ancestry of resonance.

The body (of woman) is charged with ambiguous meanings— euphoria and danger, extreme spirituality, and devilish carnality.

At the end of the nineteenth century, the metropolitan condition was marked by the spread of sexual diseases such as syphilis, and the *belle dame sans merci* of the late-romantic mythology was seen as the bearer of contagion.[20] Ascribing guilt to women here took a new turn: *la belle dame sans merci*, the beautiful pitiless woman, was the sinner who spread moral and physical contagion in the promiscuous whirlwind of metropolitan modernity.

The Frigid Sublimity of Abstraction

Symbolism was the pinnacle of romantic sublimity, but it was also the doorway to the linguistic abstraction of art and lived experience in the new century.

According to Wassily Kandinsky, "the more frightening the world becomes, the more art becomes abstract." In late modernity, the world's frightening face was in full display. According to Natalia Ilyin, abstraction was born out of the trenches of World War I. In *Chasing the Perfect*, a theoretical and historical account of the origins and meaning of modernist design, she writes the following:

> Ten million soldiers died and twenty million were wounded in the four years of "the war to end all wars," which was declared in 1914. Those numbers don't include the civilians who died, the children caught in cross fires. At the Battle of Verdun alone, a "battle" that went on for six months, 350,000 Frenchmen and 330,000 Germans died. That's about 3,778 killed a day—that's one World Trade Center a day, for six months, in one battle. [...] Imagine coming back to your nice Victorian home after that. Imagine having just lived through four years of watching your friends die hanging in the tangled barbed wire of no-man's-land. Imagine yourself, hunkered down in your trench, listening to them scream all night until the screaming stopped. Imagine coming back home after that, putting on a dinner jacket for mama's evening musical, and listening to a matronly soprano singing "The Last Rose of Summer." How were you supposed to sit on your little gold ballroom chair, wearing your dinner jacket and sipping your digestif, after what you had been through, pretending nothing had changed?[21]

"No poetry after Auschwitz," wrote Theodor Adorno after the Second World War. But this was only a way to say that empathy was dead, and that only abstract forms would be possible in the aftermath of unspeakable horror.

Just as the urge to empathy as a pre-assumption of aesthetic experience finds its gratification in the beauty of the organic, so the urge to abstraction finds its beauty in the life-denying inorganic, in the crystalline or, in general terms, in all abstract law and necessity.[22]

The history of Western civilization, and particularly the history of late modern art, may be viewed as a slow irreversible turning away from nature. The will to abstraction is simultaneously the expression of the anxiety and fear pervading the historical environment, and the condition for digital perfection.

After the sensuousness of symbolism, after the precipitation into the historical abyss of violence, the sentiment of the sublime moved towards the frigid regions of digital disembodiment. In the process of late modern abstraction, the body was denied and turned into a sanitized object. Sex was replaced by pornography and happiness by psychopharmacological maintenance. The abstract perfection of the digital world is the arrival point of this late modern trajectory: abstraction of finance from production, abstraction of work from activity, abstraction of goods from usefulness, abstraction of time from sensuousness.

Confronted with the ultimate threat, nuclear destruction and the sexually transmitted AIDS syndrome, the cyberpunk culture prepared the jump into the hyper-world of abstraction. In the cyberpunk imagination, the body was perceived as the heavy painful residual of the organic past. Cyberculture replaced the body with the sanitized, clean, smooth surface of the screen.

A sort of masculine hysteria was hidden in the digital culture of the 1980s and 1990s. Late nineteenth-century decadence originated in the spread of sexually infectious diseases such as syphilis; the techno-glamour aesthetics of the late twentieth century flourished in the aftermath of the sexual and viral epidemics of AIDS.

The prosthetic-aesthetics of the cyborg, an imaginary organism enhanced by digital prostheses, can be seen as the arrival point of romantic male hysteria that wants to escape the dangerous ambiguity of sensuousness. When romantic sublimity meets the frigid surface of digital experience, panic and depression ensue. Panic attacks are widespread symptoms in the experience of the connective generation. No longer a passionate panic resulting from the confusing, inexhaustible possibilities of nature, it is instead a frigid panic resulting from the contraction of time: a frantic time, an unattainable body, a fragmented experience, and an ever-widening space of possibilities that never become real.

The anaesthetic aesthetics of virtuality is the last avatar in a process of the sublimation of the body (of woman) that, in Christianity, stretches from Francisco de Quevedo, through Jean des Esseintes, to Georges Bataille, down to David Bowie.

In 1977, David Bowie released the single *Heroes*.

In it, he sings of a new brand of hero who emerges just in time for the neoliberal revolution, and for the digital transformation of the world.

Bowie's hero is no longer a subject, but an object, a thing, an image, a splendid fetish—a commodity soaked with desire, resurrected from beyond the squalor of its own demise.

The video-clip shows Bowie singing to himself from three simultaneous angles, with layering techniques tripling his image; not only has Bowie's hero been cloned, he has above all become an image that can be reproduced, multiplied, and copied.

Bowie's hero is no longer a larger-than-life human being carrying out exemplary and sensational exploits, and he is not even an icon.

Rather, he is a shiny product endowed with post-human beauty, an image, and nothing but an image.

This hero's immortality no longer originates in the strength to survive all possible ordeals, but in its ability to be Xeroxed, recycled, and reincarnated. Destruction will alter its form and appearance, yet its substance will be untouched. The immortality of the thing is its finitude, not its eternity. The hero is dead; long live the hero!

Videoporn and the Vanishing Body

Due to the lack of homogeneity of geo-psycho-cultures, the effects that the connective mutation has on sensitivity do not develop in a uniform way. Nevertheless a general trend seems to emerge, both in the cognitive and sexual behavior of the first generation to have grown up in a mainly connective environment.

According to the World Health Organization, suicide has notably increased, particularly among young people. Depression and panic are widespread within this generation.

Pornography is now occupying an ever-increasing space in emotional life, and religious identity has begun to fill the place that, in the context of modernity, was previously given over to social solidarity.

Is there a relation between the expanding place of pornography in emotional life and the increasing place of religion in the processes of self-identification?

According to a 1907 essay by Freud on the symptomatology of obsession and religious practices, rituals have something to do with obsession because they share the same characteristics of irrealization and compulsion.[23]

Irrealization and compulsive repetition are traits that are often found in religious behavior and in pornographic sex. Although one can find peace and well-being through religious ritual, and pleasure through pornographic consumption, both rituals and pornographic images nevertheless share the stigma of obsessive neurosis, that is, the repetition of acts that are devoid of semantic meaning and of specific efficiency.

Obsession is the compulsive repetition of a ritual that does not fulfill its aim. A ritual is a conjuration whose aim is to cohere the world of the author of the rite. From this point of view, porn has something to do with ritual. In the experience of the first video-electronic generation, porn becomes the repetition of an act of seeing that does not attain its emotional end. The ritual itself takes the place of pleasure.

Here I'm neither reclaiming any kind of original authenticity of eroticism, nor implying a supposed golden age of sexual happiness. I'm just asserting that the current proliferation of pornography is linked to an emotional pathology, highlighted by the mediatization of porn, and especially by its proliferation on the Internet. The prevailing perception of the body in the saturated infosphere is that of a place for the simulation of pleasure, which reduces the other to a projection of the mind. Since the image is separated from touch, the pornographic act, which is essentially an act of vision, does not produce the promised synesthetic pleasure. The act of vision is thus repeated over and over. The Internet is a space of endless duplication; it is therefore the ideal space for pornography. Hypertrophic stimulation, and the simulation of pleasure generate obsession.

During their long evolution, human beings have slowly learned to elaborate the stimulus of sexual excitement. The entire history of culture can be viewed as a way toward the elaboration of sexual desire. Through imagination and language, human beings have learned to balance stimuli coming from the environment with the psychic and sexual responses to it.

With the proliferation of information, the saturation of the infosphere has provoked a stimulus overload, which has an obvious cognitive effect. The time available for attention decreases.

Affective attention takes time, and cannot be shortened or speeded-up. Hyper-stimulation and visual overload are leading to a disorder in the emotional elaboration of meaning. Affective attention suffers a kind of contraction, and is forced to find ways to adapt. The organism adopts tools for simplification, and tends to smooth out its psychic response, and to repackage its affective behavior into a contracted and faster framework.

The focal point in this process of de-sensitization is the reduction of the time available for the elaboration of emotions. Pornography is one of the causes of this saturation, and one of its effects, or, better, one of its symptoms. Pornography concurs with the saturation of the infosphere, and it is simultaneously an escape from the disturbed psychosphere.

Emotion is the meeting point between body and cognition. It is the bodily elaboration of the information that reaches our mind. The time of emotionality can be fast, even very fast, or can be slow, but the elaboration of sexual emotion requires time.

Although pharmacology can accelerate sexual reactions and speed up erections, automatic engines cannot shorten the emotional time necessary for caresses. The use of sexual stimulants such as Viagra has less to do with impotence than with speed and emotional disturbances.

The electronic excitement that is conveyed through the entire mediascape puts the sensitive organism in a state of permanent electrocution. The time necessary for the linguistic elaboration of a single input is reduced as the number of inputs increases, and as the speed of input rises. Sex is spoken everywhere, but sex itself no longer speaks. Instead, it babbles, and falters, and suffers from that, or worst, ignores it. Too few words, too little time to talk, too little time to feel ... Porn can be viewed as an attempt at emotional automation and at creating uniformity in the temporality of emotional response.

Reformatting the mental sphere implies reformatting the emotional dimension. The new format is smooth, connectable, and recombinant. The hair-free slickness of the digital image has invaded the sexual imagination, and caused it to mutate. Experience of videoporn expands as the physical presence of the body of the other grows increasingly rare.

Similarly, we can say that religious belonging becomes progressively more important in the life of postmodern populations as the sense of belonging to a territory, a community, or a social class is lost. The current resurgence of religious forms of self-identification has little to do with spirituality, piousness, or the feeling of the sacred. On the contrary, it has a lot to do with the craving for belonging, which seems to accompany the deterritorialization and the decrease in solidarity provoked by virtualization.

Russian Spiritualism and the Wreckage of Communism

In an essay entitled "The Revelation about Man in the Creativity of Dostoevsky," Nikolaj Alexandrovic Berdyaev evokes the "metaphysical hysteria of the Russian spirit."[24] Based on the ethical and aesthetic exaltation of unhappiness, this hysteria leads to a kind of mysticism based on suffering.

Russian literature can be considered an extreme manifestation of theatrical, romantic self-immolation. To understand the peculiar vibration of Russian subjectivity, calls for an analysis of the cultural and theological specificity of Eastern Christianity. Its difference from

Western Christianity is essentially based on the spiritualization of the person of Christ, whose carnality is denied or at least eclipsed. If Plotinus was ashamed of having a body, Russian culture seems to be haunted by the idea of corporeity as execration, as a crime finding penance in itself.

It seems that having a body is a source of continuous, unavoidable pain, and that pain is the deserved punishment for the guilt of having been born in that horrible place called Russia. The rhetoric about the beauty of the homeland and of the Russian people can be read upside down, as a manifestation of the masochist Orthodox taste for self-inflicted pain. Russian writers chant the infinite pain that life inflicts on living beings, and this pain they call joy.

Theological questions concerning Christ's incarnate humanity— the core of the debate about the nature of Christ—are crucial for the genealogy of Orthodox culture. Herein lies the exacerbation of spirituality, and of the search for purification that runs throughout the history of Eastern Christianity, taking historical and political form in Russian subjectivity. In the cultural sphere of Eastern Christianity, human existence is considered a condition of absolute alienation, a source of perpetual unhappiness.

The Italian Slavist Vittorio Strada writes: "Western Christianity acts in the space of History, Eastern Christianity strives for Eternity."[25]

Desire for the absolute is visible in the historical sphere as a will to total palingenesis and to the purification of the social community from traces of the past. This reference to purity is quite transparent in the Russian conception of revolution, particularly in the Leninist persuasion that the revolutionary political party was the "incarnation" of the pure idea descendent from German philosophy, and that it must be embodied by a small organization of professional bearers of revolutionary truth.

The Russian exacerbation of the role of pure subjectivity entered the scene of world history in 1917. The Soviet Revolution—which Lenin managed to unleash against the will of many prominent leaders of the Russian socialist movement—provoked a catastrophic polarization of worldwide social conflict, and forced the workers' movement to identify with a totalitarian experiment based on a class struggle structured by an authoritarian state.

The Russian revolution provoked an irreversible rupture and a permanent laceration in the body of society whose effects persisted worldwide throughout the century. Lenin forced the workers of the world to defend the socialist state of the soviets, and to enter a process of permanent war. This war lasted until 1989, but the class of workers had been doomed to inescapable defeat since the beginning.

Messianic utopianism, widespread in Russian society during the nineteenth century, merged with an extremely voluntaristic Bolshevik project so that the history of the communist revolution began as a tragedy, in a context of cultural immolation, and was destined from the beginning to end tragically as well. The violence and authoritarianism that the Leninist experiment unleashed throughout the country, and exported worldwide, brutally changing the prospects of the international movement for workers' emancipation, had been inscribed in Russian history during previous centuries, and still prevail in Russian political life after the end of the soviet dictatorship.

Bolshevism and Psychic Depression

Leninist communism may have ruined Russia. But Russia has (forever?) ruined communism as a possible alternative to capitalism. Indeed, Russian subjectivism and the cult of purity dragged the international workers' movement into a vision of permanent military mobilization that was not part of the Marxian schema.

The Leninist conception of communism has little to do with Marx's thought, and with the history of the workers' struggle against capitalist exploitation. It has much more to do with the history of chiliastic sectarianism and of the dualistic, Manichean spiritualism that has permeated Russian culture since the time of Bogomil's influence on the Orthodox Church.

According to Nikolaj Alexandrovic Berdyaev:

> The Russian people did not achieve their ancient dream of Moscow, the Third Rome. The ecclesiastical schism of the seventeenth century revealed that the muscovite tsardom was not the third Rome. The messianic idea of the Russian people assumed either an apocalyptic form or a revolutionary one; and then there occurred an amazing event in the destiny of the Russian people. Instead of the Third Rome in Russia, the Third International was achieved, and many of the features of the Third Rome passed over to the Third International. The Third International was also a Holy Empire, and it was also founded on an Orthodox faith. The Third International was not international, but a Russian national idea.[26]

The spontaneous goal of the workers' movement was to expand the space of autonomy from capitalist exploitation. The idea that the movement was caught in a dialectical contradiction was an effect of the Hegelian interpretation of social process. This idea became historical reality when the Russian palingenetic cult of purity blended with the Hegelian tradition.

In my opinion, the fusion of Marxism and Leninism was the origin of the workers' defeat. Lenin brought to the workers' political discourse an element of subjectivism and of purity that did not belong to the experience of autonomous social movements. The workers' movement aimed at emancipating life and territory from capitalist domination, but the Leninist breakthrough transformed the movement into a project of absolute separation from existing reality, a project of radical demolition, and of palingenetic purification.

Based on the denial of Christ's human embodiment, the distinctive particularity of Russian spiritualist hysteria is purity. Lenin inherited this cult of purity, and built his revolutionary party on its premise.

The fundamental text for Leninist politics is *What Is to Be Done?*, published in 1902. In this text, the revolutionary party is described as a collective intellectual pursuing a project that does not depend on the concrete history of social struggles.

In the first part of the text, Lenin confirms Lassalle's argument that purification, that is, epuration and cleansing, is strengthening the party. This idea of ideological cleansing constitutes the main thread in the history of the Soviet Union's Communist Party, particularly during Stalinism.

According to Lenin, the working class in its spontaneity could only develop an economic trade-unionist action, but was unable to develop a process of political revolution.

This impurity of the working class had to be overcome, so that society could adapt to the purity of the communist ideal. Only a party that was the bearer of the pure Logos, rather than the simple aggregation of impure social bodies, could be the bearer of the revolutionary project.

If we look at the reality of social conflict, we understand that it has nothing to do with the purity of ideas. The workers' struggle expresses a radical refusal of exploitation, but also the ability to coexist and coevolve with the capitalist machine. Thanks to such contamination, the workers' struggle can simultaneously be a practical critique of political economy and a dynamic engine of industrial development. It has deployed itself in a space that is autonomous from capitalist rationale, without breaking the link with technical and political innovation. Social conflict has simultaneously acted to disrupt the process of production, as well as to cause innovation and transformation.

I think that it is important to link the catastrophic Russian revolution of 1917 with the intellectual and psychological biography of Vladimir Ilyich Ulianov, otherwise known as Lenin.

In her biographical essay on Lenin, Hélène Carrère d'Encausse, a French historian of Georgian origin, mentions two major episodes of clinical depression that are generally ignored by the hagiographic Leninist tradition. Her book is particularly interesting as it focuses on the emotional life of the communist leader, the importance of his relation with his mother, his sister, and in particular with his wife, Nadezhda Krupskaja, who took care of him during periods of acute psychic crisis. The book also discusses Inès Armand, the lover who intruded into Lenin's life, and who was later removed and neutralized as a potential danger to the leader's political integrity.

Depression is the striking feature of Lenin's psychological portrait. Crises of depression coincide with the most important political decisions of his life. The first major crisis, according to Hélène Carrère d'Encausse, occurred in 1902, concurrent with the decision to found the Communist party and draft *What Is to Be Done?*. The second occurred in 1914, when Lenin decided to break with the Second International before the Zimmerwald Conference and the communist schism throughout Europe. The third happened in spring 1917, simultaneous with the decision to launch the soviet insurrection that took place in October.[27]

These decisions marked the emergence of communist identity, and imposed a voluntary acceleration of the history of the class struggle throughout Europe and the world. In my opinion, they can be linked to Lenin's cycle of depression. When intelligence is depressive, will is the only therapy that makes it possible to ignore the abyss. The abyss is not removed, resolved, avoided, or overcome. It is ignored, but it remains, and the decades following the revolution exposed its persistence, leading the century down into the abyss.

More than the political meaning of Lenin's decisions, I am interested here in the relation between Bolshevik voluntarism and a masculine inability to deal with depression.

From the political point of view, the Bolshevik breach precipitated the confrontation between workers and capital worldwide. Workers were pushed towards a totalizing form of opposition, and towards civil war. Social autonomy was forced to choose between revolutionary terror and capitulation. And where communist parties succeeded in seizing political power, they turned to violent dictatorships and the submission of social life. This is how the Leninist strategy prepared the worldwide catastrophe that, at the end of the twentieth century, led to the worst of all possible defeats, and whose effects we'll be experiencing for decades.

The project of worldwide renewal, beginning with the palingenetic violence of the revolution, is a mythology without any historical foundation. History has never known abolition, palingenesis, or rebirth. History is always about stratification, negotiation, coevolution, autonomy or dependence, identification or *extraneousness*.[28] It is not about abolition.

Leninism can be considered an attempt to deny depression, an assertion of the purity of the will, and a refusal to accept the finitude of human potency. It can be seen as the masculine hysteria that was already at work in Dostoyevsky's writings.

According to Berdyaev, writing in 1915, "in the Russian character there is a certain metaphysical hysteria, which Dostoevsky so powerfully sensed and revealed."[29]

The hysteria that Berdyaev is writing about can be seen as a mystical euphoria for the unhappiness that nature and human actions inflict on the human body.

"Mother [...] don't cry, life is paradise, and we are all in paradise, though we don't want to acknowledge it," writes Dostoyevsky in *The Brothers Karamazov*, with a sort of ironic consciousness of the hysterical feature of his narrative imagination.

In the nineteenth century, psychiatrists saw hysteria as an essentially feminine affection, linked to the negation of the sexual body. Male hysteria concealed itself behind high ethical values and spiritual excitement. But the truth was the same: the terror of being in contact with the body, which is always the body of the other, because one's own body is other in relation to the inner self. In Tolstoy's *Kreutzer Sonata*, as in Chekhov's *A Dreary Story*, the male fear of the woman's body is provoked by the cultural inability to imagine love as mutual availability and ironic game, and because of the dramatization of touch, love is intrinsically viewed as metaphysical guilt. The unbridgeable distance between the sphere of spirituality and the sphere of physical pleasure feeds the hysterical inability to experience joy in life.

Japanese Dis-Identification

Moving back and forth in the labyrinth of late modernity, I have so far tried to establish the connection between psycho-cultures and sensibility, and also between sensibility and historical processes of subjectivation. The mutation underway in the new century can be linked to the technological transition to the digital environment, but this transition takes different forms and coheres in different ways according to different cultural environments. It is for this reason that I am trying to trace different pathways through this cultural mutation according to different psycho-anthropological configurations anchored in their geo-cultural locations.

A privileged point of view for the investigation of the current mutation is Japan, since the dynamic relation between the obliteration and the persistence of a denied cultural past has particularly affected the Japanese experience of modernity.

According to the psychoanalyst Takeo Doi, the marking feature of Japanese psychology is *amae*, a concept that designates a passive form of love, or better, a form of trust and self-committing.[30]

Takeo Doi describes *amae* as the emotion experienced by the unweaned child. But *amae* may also be seen as a sort of indulgent dependency based on the attribution of a reassuring symbolic function to the other, for example, the father, the husband, the political authority or the corporation. Such an attribution of symbolic function places the other in a position of authority and certainty.

In articles, interviews, and talks delivered during his travels to Japan, Félix Guattari sketched out a number of interesting anthropological considerations about Japanese subjectivity. Some of his reflections correspond to Takeo Doi's identification of *amae* as the marking feature of Japanese psycho-culture.

Guattari was fascinated by the symbiotic genesis of Japanese subjectivity, by the openness and the prehensile disposition of the Japanese mind. In the discussions that he had with the Japanese philosophers and intellectuals Masaaki Sugimura, Asada Akira and

Shin Takamatsu, Guattari seemed to be interested in how technological modernism and archaic cultural traits are creatively combined in Japan.

Despite the persisting myth of ethnic homogeneity, the reality of Japanese culture, and particularly of Japanese language, is one of openness and permeability.[31] Religious faiths are often blended, and the Japanese language is a mix of Chinese influences and of Western borrowings. This adaptability first manifested with the assimilation of Chan Buddhism coming from China. Confucianism influenced the formation of the leading political class, and easily coexisted with the original indigenous Shintoism.

> Ideas are for the Japanese nothing more than tools that can be used for various purposes. If a saw does not do the job, you can use an axe. In the same way, if Confucianism does not give the desired result, resort may be had to Buddhism.[32]

In the second part of the nineteenth century, the political crisis affecting the Shogunate opened the way to the so-called Meiji Restoration, which was not in fact a restoration, but the establishment of a process of modernization based on importing Western techniques and procedures.

The educational system was reformed according to the suggestions of Prussian consultants, and the industrial system was created and developed in collaboration with British and American experts. Even the constitution of the modern Meiji state was written along the lines of the Prussian constitution. The history of the so-called Restoration—which was in fact the creation of a new political reality specifically inspired by Western experiences and institutions—was the artificial reinvention of the past according to a political and cultural project that was completely oriented towards the future. The past was forgotten and obliterated, since Japanese identity seemed definable as a process of continuous self-definition.

Nervous Breakdown

In *Hakuchi*, a film based on a short novel by Ango Sagacuchi, the director Makoto Tezuka imagined that the nuclear bomb had not been dropped on Japan, and that war had continued for twenty or thirty years. Madness is at the center of the movie: the madness of those who have been forced to live their entire lives under conditions of constant stress.

Since the end of the nineteenth century, Japan's enforced modernization strained the nervous energies of a people who were deeply immersed in a traditional way of life.

Natsume Soseki, author of *I'm a Cat* and *Sanshiro* among many other works, is considered the most brilliant novelist of his time. At the beginning of the nineteenth century, he predicted that Japan was heading for a collective nervous breakdown because of the attempt to quickly digest Western civilization and the rhythm of modernity.

While in London during a journey to Europe, Soseki experienced the suffering and stress provoked by the jarring rhythms of modern Western life, and wrote the following:

> The two years I lived in London were the unhappiest two years of my life. Among the English gentlemen, I was like a lone shaggy dog mixed in with a pack of wolves.[33]

The sentiment of being like a lone shaggy dog among a pack of wolves is extraordinarily touching, but also conveys the sentiment of humiliation that the mutation imposed by the Meiji reformers provoked in many intellectuals, and, in different ways, on the population at large.

In my opinion, a distinctive aspect of the psychological violence intrinsic to enforced modernization lies in the forced masculinization of Japanese psychology. This resembles in many ways the violent de-feminization enforced by Italian Futurism on the psycho-political perception of self of the Italian people during the Fascist revolution.

Italian poetry and art had always portrayed their Mediterranean country as a beautiful woman.

"Oh, my own Italy, though words be useless/To heal the mortal wound/I see covering all your lovely body," writes Petrarch in the poem "Italia Mia," identifying Italy with a beautiful feminine body.[34]

In "The Futurist Manifesto," F.T. Marinetti expressed "the scorn of woman," intending to emphasize the manly values of the nation facing economic growth, colonial wars, and international competition.

In a similar vein, the Japanese were compelled to strengthen their character and deny their feminine kindness and timidity when forced to become a colonial power and to fight modern wars so as to compete on the world stage. Despite the rhetoric of honor embodied in the tradition of the samurai, traditional Japanese culture is marked by self-restraint and apprehensiveness. Stress and psychological violence toward the self, resulting from the repression of spontaneous emotionality during the modernization process, led to the collective nervous breakdown that Natsume Soseki had warned about, and which became evident in the years of massive conversion to tenno-psychopathic ideology, a hysterical version of Hitlerian political aggressiveness.[35]

In *Bodies of Memory*, Yoshikuni Igarashi describes the most dramatic moment of the fascist Japanese adventure, the defeat, the surrender, and the meeting of Emperor Hirohito with General MacArthur.

Immediately after the defeat of Japan, the United States and Japan recast their relationship in terms of a melodrama of rescue and conversion. According to this melodrama, the United States rescues a good enemy, Hirohito, from the deleterious elements in the enemy country, and the good enemy becomes converted into a representative of US values. Hirohito emerges as a desirable object in the drama to explain why he deserves to be rescued; both countries' relations are expressed through a drama that features an entanglement of desire for the other.

The relationship between the United States and Japan in the postwar melodrama is highly sexualized. The drama casts the United States as a male and Hirohito and Japan as a docile female, who unconditionally accepts the United States' desire for self-assurance.[36]

In his journal, MacArthur describes the scene of his meeting with Hirohito in an overly theatrical fashion that speaks volumes about this sexualization, and underlines the feminine nature of the Japanese psycho-cultural condition.

He was nervous and the stress of the past months showed plainly. I dismissed everyone but his own interpreter, and we sat down before an open fire at one end of the long reception hall. I offered him an American cigarette, which he took with thanks. I noticed how his hands shook as I lighted it for him. I tried to make it as easy for him as I could, but I knew how deep and dreadful must be his agony of humiliation.[37]

After the defeat, and the conversion to Western economic values, competition shifted from the field of military aggression to that of productivity and economic growth. There is no doubt that the Japanese performance was highly effective in the post-war decades, but the price it required was similarly high. It is thus possible to assert that contemporary Japan is a sort of laboratory of stress-related psychopathology, and in particular, a laboratory of the pathologies related to the connective mutation of technology and of social behavior.

The suicide of writer Yukio Mishima, in 1970, rang an alarm bell signaling the psychic suffering provoked by the enforced homologation of Japanese culture to Western values. Mishima was certainly nostalgic for the past military tradition of his country, but what is more interesting in his last words, and in his hara-kiri, is the perception of an unbearable cognitive dissonance which grows increasingly painful, notwithstanding the triumphal successes of Japanese capitalism.

Thanks to the systematic cancellation of memory, and to the rewriting of collective identity, the smooth surface of social life becomes perfectly suited for increasing productivity and ever-growing consumption. Social language is purified from any historical incrustation and from the impurities of emotional life, such that the economy becomes the universal language.

Once more, the puritanical denial of the past acts as an introduction to the process of virtualization. Cleared from the superfluous fuzz of cultural identity and psychic emotion, the smooth surface of social life becomes perfectly compatible with the digital system of exchange. The post-human mutation deploys itself more efficiently where dusty memories are cancelled.

Hikikomori

In 1977, at the end of summer vacation, a wave of child suicides caused a general outcry throughout the country. Within a few days, thirteen primary school children had killed themselves. The gratuitousness and incomprehensibility of the gesture was particularly disconcerting, since there seemed to be no motivations or reasons for the act. There was a striking lack of words, an inability on the part of the adults who lived with the children to predict, understand, or explain what had happened.

In 2002, the filmmaker and poet Sion Sono wrote and directed the movie *Suicide Club*, which featured a massive wave of apparently unconnected suicides committed by young students.

A suicide epidemic seemed to grip the country.

In 1983, a group of students in a secondary school murdered some elderly, homeless people in a park in Yokohama. When questioned, the children offered no explanation other than that the homeless people they killed were *obutsu*, dirty and impure things. As in the *manga* comics that achieved mass readership precisely in the second half of the seventies, the enemy was not evil, but dirtiness. Cleanliness, and ridding the world of the "waste products" of the indefinite, the confused, the hairy, or the dusty, prepared the way for digital, smooth surfaces without asperities. Erotic seduction was progressively disconnected from sexual contact, until it became sheer aesthetic stimulation. It was in Japan that the first symptoms of this trend were spotted.

In addition, since the early years of the new century, a new behavior began to spread in Japan. A number of people, most of them young, break their ties with the outside world, and lock themselves in their rooms, refusing any kind of socialization except for a virtual connection with others of the same ilk. They have become so numerous that they are officially labeled *hikikomori*, and the Japanese state has been forced to take their numbers into account, and to provide social aid for those among them who need assistance.

According to the official definition, hikikomori are people who live in a state of complete isolation. Researchers have suggested six specific criteria required to *diagnose* the hikikomori condition:

1) most of the day and nearly every day is spent confined to the home,

2) a marked and persistent avoidance of social situations,

3) symptoms that interfere significantly with the person's normal routine, occupational or academic functioning, or social activities or relationships,

4) withdrawal is perceived as ego-syntonic,

5) lasting at least six months,

6) no other mental disorder that can account for social withdrawal and avoidance.

According to the estimates of the Japanese Ministry of Health, in 2010, 700,000 people averaging around 31 years of age were living as hikikomori, and 1.5 million people were estimated by the government to be on the verge of becoming hikikomori.

Psychiatric definitions of hikikomori culture seem to me to be quite elusive, since they do not address the crucial problem that is implied in the behavior of so many Japanese kids. Such behavior should not be seen simply as the symptom of pathology, but should be understood as a form of adjustment to the anthropological and social mutation that is underway, as an answer to the unbearable stress of competition, mental exploitation, and precarity. According to Michael Zielenziger's 2009 book, most of the hikikomori he interviewed showed independent thinking and a sense of self that the current Japanese environment could not accommodate for. Considering the stress of competition provoked by the Japanese economic context, one may argue that hikikomori behavior is a healthy reaction to the frantic, precarious life created by late capitalism: a fully understandable withdrawal from hell.

The Dark Side of Superflat

In the last decades of the past century, Japanese financial power, desperate to find a new, international image for Japan, exploited artists as a resource to export the brand of *Cool Japan*. The time of a fast-growing economy was over, and stagnation looked like the long-lasting condition of an ageing Japanese society. It was at this time that Superflat art came to dominate the world's view of Japanese contemporary art, monopolizing spaces, and offering a distorted vision of the life of the young: cute schoolgirls, Gothic Lolitas, and sweet colorful gadgets.

Cool Britannia had been the trademark of the post-Thatcher age, a time when Young British Artists managed to reinvent London, hiding the reality of the aggression against the welfare state, the impoverishment of large areas of the country, and the exploitation of precarious young workers. Young Japanese artists—Takashi Murakami, Yoshitomo Nara and Mariko Mori among others—replicated the process, invented Cool Japan, and exported Superflat art to the West. They joined forces with art world entrepreneurs, fashionable gallerists, foreign dealers, and curators who helped them to create the image of Cool Japan for the world market. Large financial groups such as Mori Building Co. and political leaders from the conservative Liberal Democratic Party appropriated the creative surge and the glamour coming from the Superflat wave.

With American pop as its antecedent, [...] Japanese "neo-pop" paro-
died the infantilism of post-war Japanese consumer culture, making
art by "sampling" and "remixing" the endless array of consumer junk
with which Japanese filled their pacified US-dependent lives.[38]

Otaku is the Japanese version of the international nerd, who tries to
embody the positive values of the creative class, simultaneously
suffering the stress and affective misery that is part of the competitive
game. Superflat was an attempt to translate the otaku style for an
American mass audience, and this worked at the level of the market.
My Lonesome Cowboy, Takeshi Murakami's pop sculpture, sold for
13.5 million dollars, and Mariko Mori's cute, glamorous high-tech
gadgetry went mainstream in the global video-scape.

Mariko Mori trapped herself in a classic *neo-Japoniste* trope:
Japan as futuristic techno-paradise. There she was: the pretty girl in
an astronaut's jump suit welcoming the tired space-traveler with a
few words of mystical Asian religion. It was contemporary Japan as
it had been envisaged in the Osaka World Expo of 1970.[39]

But according to Adrian Favell,

the easy eye candy of superflat art was, to anyone that knew
anything about the place, a blatant caricature and distortion of
modern Japan. For a decade, it became practically the *only* Japanese
contemporary art *ever* seen internationally. In fact, the success of
their *otaku* style art stood as the stunning exception to the dismal
failure of much Japanese contemporary art to match the international impact of Japan's other creative industries.[40]

Superflat did not help to explain the psycho-cultural reality of the country, particularly the reality of the precarious generation, "a generation

who," according to Favell, "had never held a proper job, never had to deal with a real relationship, lived with its parents, and sat day and night at home poring over its obsessively catalogued collections."[41]

As Japan's postwar boom years came to an end at the beginning of the 1990s, the country entered a period of long, slow decline that has continued through to the environmental disaster of 2011.

The year of Fukushima marked a change in the perception of Japanese society, haunted by depression, loneliness, and suicide.

Superflat aesthetics emphasized the glamorous side of the smooth sensibility emerging from the process of digitalization. But the glamour has dissolved, and what is left is the pathological side of smooth sensibility: the lonely depression of the nerd.

Connectivity and Suicide in South Korea

As I'm trying to retrace the anthropological and psycho-cognitive effects of the connective transition underway, I cannot fail to mention South Korea, the country with the highest rate of broadband connectivity in the world.

Just as it was in Japan and possibly even more so, the first part of the twentieth century in South Korea was marked by a continuous state of war, violence, devastation, and bombing. The Japanese colonization had aimed to cancel national identity, and two wars had physically destroyed most of its urban territory. Finally, in the last decades of the century, an intense process of industrialization and territorial redesign had erased those parts of the natural landscape that had not been destroyed by war.

In the South Korean capital city of Seoul, traces of traditional life have been overwhelmed by artificially redesigned life. Social communication has been thoroughly transformed by mobile smartphones that bring the network into daily life and into urban wanderings. The majority of people are always looking at their small cell-phone screens. In the land of Samsung and LG Corporation, connection is permanent. While walking, sitting in a coffee shop, standing, or waiting for the subway, the hands of city dwellers are busy fingering their screens.

Urban vision has been thoroughly redesigned by screens of all sizes placed everywhere: big screens on the walls of skyscrapers, middle-sized screens in the halls of train stations. The small private smart-phone screens prevail over the attention of the crowd as it calmly and silently shuffles about, seldom looking around. The South Korean mind has been reshaped in this artificial landscape, and has smoothly entered the digital sphere with a low degree of cultural resistance compared with other populations of the world.

In a cultural space emptied by military and cultural aggression, the Korean experience is marked by an extreme degree of individualization, and is simultaneously headed towards a thorough wiring of the collective mind. The individual has become a smiling monad walking alone in urban space, in tender, continuous interaction with the photos, the twitter feeds, and the games pouring out from his or her small screen. As social relations are mediated by connection, its rules and procedures are hidden in the technical format of the Net.

Perfectly insulated and perfectly wired, the organism has become a smooth interface of the flow. In order to access interaction, the individual has to adapt to the format, and his/her enunciations have to be compatible with the code.

Korean cultural history seems to be particularly suited to this transfer of social life into digital format. Hangeul, the Korean alphabet invented in the fifteenth century by King Sejong, seems to be one of the reasons for the late modern economic success of the country. In fact, according to many linguists and anthropologists, the Korean ability to transmit digital content faster than in any other country of the world has something to do with the Hangeul writing system, which is ideally suited for digital technology.

> Korea is a world leader in cell phone technology and production, while the speed with which users can exchange text messages via cell phone keypads is unquestionably the fastest in the world. Like the Hangeul alphabet, Korean cell phone keypads are based on a principle of adding strokes to basic consonants and vowels, which means that a minimal number of keys are needed to create the entire alphabet.[42]

Within the span of a single working life, the Korean economy has grown seventeen-fold, but the prosperity Koreans enjoy has not relieved the competitive pressure they endure. In the space of two generations, South Koreans have moved from starvation to advanced levels of consumption, but the last generation is caught in the trap of labor precarity, competition, and social anxiety. Young people are increasingly obliged to take on debt if they want to study, marry, and rent a house.

South Korea has the highest rate of connectivity, and one of the highest suicide rates in the world. Among the OECD countries, Korea leads the gloomy contest with 28.4 suicides per 100,000 people. Second comes Hungary with 17, then Finland, and Japan.

Even if I want to avoid deterministic causation, I must underline this significant point: three of these highly suicide-prone countries (Japan, Finland and South Korea) also have high connectivity rates. Is there a link between high connectivity and suicide?

As a result of my research on the psychological effects of the technological evolution, the answer is yes. There is a link between connectivity and social proxemics; there is a link between connectivity and loss of empathy; there is a link between connectivity, precarization of labor, and loss of solidarity. There is a link between connectivity and suicide.

Suicide is the most common cause of death for those under 40 in South Korea. Most interestingly, the number of suicide deaths in South Korea doubled over the last decade, and quadrupled during the three decades of the electronic changeover, from 6.8 per 100,000 people in 1982 to 28.4 in 2011.

Over the span of two generations, living conditions certainly improved in terms of revenue, nutrition, freedom, and the possibility to travel abroad. But the price of such improvements has been the desertification of daily life, hyper-acceleration of rhythms, the extreme individualization of biographies, ferocious competition in schools and in the labor market, and a precariousness of work that also implies unbridled competition. This undoubtedly explains the extraordinary propensity of Koreans both young and middle aged to commit suicide.

High-tech capitalism implies improvements in revenue and consumption, but also implies ever-increasing productivity, constant competition, and an endless intensification of the rhythms of work. Koreans look back at the living condition of their grandparents, and the impressive improvements in economic conditions obscure their present alienation. However, for some, this alienation is unbearable. The desertification of the landscape and the virtualization of emotional life are converging into feelings of loneliness and despair that are socially difficult to consciously refuse and oppose. Loneliness, stress, competition, a sense of meaninglessness, compulsion, and failure: Yearly, 28 people out of 100,000 succeed in their escape attempt, although many more try, unsuccessfully, to escape.

3

The Aesthetic Genealogy of Globalization

> When we reflect on the modern age, we inquire after the modern
> world picture. [...] "world picture" does not mean "picture of the
> world" but rather, the world grasped as picture. [...] The funda-
> mental event of modernity is the conquest of the world as picture.
> From now on the word "picture" means: the collective image of
> representing production.
> —Martin Heidegger, *The Question Concerning Technology and*
>   *Other Essays*

The idea of writing something about the transition from the conjunctive to the connective dimension of sensibility came to my
mind in the spring of 1997 after two successive journeys, one to
Cordova, Andalusia, the second to Jaipur, Rajahstan. In Cordova I
visited the Great Mosque, and in Jaipur I visited the Amer Palace.
Both places are remarkable for the mingling and overlapping of the
Islamic style of abstract decoration with elements coming from the
figurative styles of Catholic and Hindu visual culture. Cordova and
Jaipur have been border cities for centuries.

It was then that I started to think about the points of overlap
and transition in aesthetic sensibility and aesthetic production.

At that time, cyberculture and the new economy were at their pinnacle, animating new prospects of development and civilization. Notwithstanding these optimistic expectations of expansion, prosperity, and peace, in the underground, the sound of the old, modern world imploding was audible. Fundamentalism was sweeping through Algeria, civil war was destroying Yugoslavia, and oppression of women in a Taliban Afghanistan financed by the US was systematic. The millennium of bourgeois civilization was approaching its end while a new barbarianism was visible in the distance. It had taken a thousand years to turn the barbarian into a bourgeois, but only a few decades to turn the bourgeois into a barbarian.

After my journeys in Andalusia and Rajahstan, I focused my attention on the psychopathology of desire. In the second half of the decade, the first post-alphabetic generation was emerging on the world scene. This was a generation that had been exposed since birth to the electronic flow of synthetic images, and whose perception had formed in an environment where most experiences were the effect of simulation. Involved in a process of mutation, this generation was developing new cognitive competences while others atrophied.

Over the course of human evolution, the spheres of eroticism and of aesthetics have always been linked, and their contamination has generally emphasized erotic pleasure while enriching aesthetic sensibility. At a certain point in late modernity, when proliferating information had invaded the social organism, aesthetics and eroticism entered a zone of disturbance. There, the conscious organism confused the access codes, exchanging aesthetic and erotic signs such that it lost itself in a labyrinth.

More than fifteen years have gone by, and my work seems unending. Since the focus of my research is a mutation that is currently underway, its territory expands while I attempt to map it and its scenario continuously changes, jeopardizing my provisional systematizations and interpretations, and continuously displacing my point of view. I've thus given up any pretensions to being systematic, and am simply attempting to map local, particular objects such as artworks, novels, movies, or events that can be considered examples of the mutation.

While connection is introducing pathological contractions into the flow of conjunction, it is simultaneously opening a new horizon for communication. Is it possible to imagine a new, happy balance of sensibility and sensitivity, a new post-natural nature of the human sensorium? Or should we instead imagine lines of escape from the spreading universe of unhappiness, islands of slowness and convivial corporeality?

Making the World Invisible: Capitalism and Semiocapitalism

I call semiocapitalism the present configuration of the relation between language and the economy. In this configuration, the production of any kind of goods, whether material or immaterial, can be translated into the combination and recombination of information (algorithms, figures, digital differences).

The semiotization of social production and economic exchange implies a deep transformation in processes of subjectivation. The infosphere acts directly on the nervous system of society, affecting the psychosphere, and affecting sensibility in particular. For this reason, the relation between economics and aesthetics is crucial to an understanding of today's cultural transformation.[1]

During the last decade of the past century, the info-econosphere emerged as a unified field of research for the social sciences.

Economics and media sociology underwent a process of integration, and many authors underlined the relation between the digital economy and socio-cultural transformations.

Understanding the historical and anthropological processes that have subjugated modern social activity to economic reason is a task for the social sciences, but I believe it is also a task for aesthetic theory.

In his seminar on the birth of bio-politics Foucault linked the genesis of economic reason to the history of capitalist domination, and spoke in particular of the universal uniformization of human culture in the operational model of the *enterprise* as the core neoliberal project. Foucault described capitalism as the subjugation of social potency (physical and intellectual) to the rule of scarcity and accumulation, and saw economics as the ideological technique used to transform use value into abstract value, to erase the concrete reality of human activity, and to emphasize the abstract function of labor, that is, to create surplus value which would make accumulation possible. The potency of labor is forced to bend to rules of exchange that are not natural laws but social rules, effects of a political decision and of a language act that marks the social body.

According to Foucault, the genesis of economic reason can also be described from the point of view of the modeling of sensibility.

Economics is a special way of making things visible.

In Plato, the sphere of the visible is despised and condemned on behalf of a higher sphere, the sphere of pure understanding and of spiritual intellection. A wide iconoclastic tradition marked the history of monotheistic religions.

Modern economic reason plays an ambiguous role in terms of the ancient dichotomy between materialism and idealism—which has something to do with the dichotomy between the visible and the invisible.

In its industrial phase, capitalist economy assessed the dignity of visible materiality, of the physical transformation of material stuff, and of the bodily enjoyment of material things. The creation of indust-reality was linked to the secularization of the visible.

During the transition to modern civilization, the shift from a Renaissance perspective to the Baroque proliferation of perspectival viewpoints conspired to establish economic reason in the space of visibility and of the image. During the centuries of modern evolution, indust-reality produced a change in the field of visibility. The physical world was modeled and transformed according to rational projects for the urban landscape, and according to the standards of mass manufacturing.

In its most recent phase, capitalist production marginalized the physical transformation of matter and the physical manufacturing of industrial goods, enabling capital accumulation through the recombination of information and the manipulation of financial abstraction.

In the sphere of capitalist production, the visible, material physicality of use value is but an introduction to the holy abstraction of exchange value. This is the double movement that Marx called *commodity fetishism*.

The process of invisibilization of the world is at the core of the abstraction process, the main trend of the relationship between the real world and the economy. "All that is solid melts in the air," wrote Marx in *The Communist Manifesto* in 1848. But in indust-reality, the invisible goal of abstract valorization is obtained by the physical manipulation of visible things.

Semiocapitalism dissolves the visible process of production. Financial capitalism is the utter dissolution of the sphere of visibility and the melting of capital accumulation into the abstract kingdom of virtual exchange.

Global Imagination and the Baroque

While the territorialized economy of the bourgeoisie was based on the iconophobic severity of iron and steel, today, the economy is based on the kaleidoscopic, deterritorialized machine of semiotic production: those commodities that circulate in economic space are signs, figures, images, projections, and expectations.

No longer a mere tool for the representation of the economic process, language becomes the main source of accumulation, constantly deterritorializing the field of exchange. Speculation and spectacle intermingle because of the intrinsic inflationary (metaphoric) nature of language. The linguistic web of semio-production is a game of mirrors that is inevitably disturbed by crises of over-production, bubbles, and bursts.

To get a general sense of our postindustrial cultural transformation, it is useful to reassess the role that the baroque played in modernity, and to understand how it is now operating behind the scenes of our contemporary imagination and perception.

According to Deleuze, the baroque is a transition that involves perception, imagination, and the social environment.[2]

Let's go back to the golden age of the baroque.

The discovery of the New World and the colonization of the Americas marked the beginning of modern expansion. The sudden change and enhancement of the universe of experience was the source of the baroque spirit, a spirit based on the idea of the infinity of God's creation. Beginning with the golden centuries of the Spanish expansion, the baroque sensibility opened the door to the ontology of infinite proliferation and therefore to the experience of modernity.

In the seventeenth century, at the highlight of the baroque period, the infosphere underwent an impressive mutation. The discovery of new lands provoked a distorting expansion of experience, and printing technology made it possible to distribute written text. What had previously been a rare and expensive source of privilege, the book and written text in general, spread largely throughout Europe in the sixteenth century. The dissemination of copies and reproductions, combined with the propagation of perspective in painting, paved the way to the simulation of imaginary words that would invade the spheres of daily experience and social culture.

In the sixteenth century, inflation disturbed the order of the economy for the first time. This new economic phenomenon was recorded as an upsetting event affecting exchange, as well as psychic and linguistic spheres.[3]

Spanish poets such as Francisco de Quevedo and Luis de Góngora refer to *locura* (madness) as an effect of the inflation of meaning that was provoked by the proliferation of semiotic stimulation. Semiotic inflation went hand in hand with monetary inflation.

In the lexicon of economics, inflation means that a greater amount of money purchases less goods. Similarly, semiotic inflation implies that more and more signs purchase less and less meaning. Semiotic inflation can be described as an excess of signs overwhelming conscious attention until it ruptures the link between sign and referent.

The baroque is the recording of this *exorbitation*: this excess of signs and the randomized relation between meaning and signs.

The phenomenon of fashion (*la mode*) and the infinite game of the appearance of things such as clothing, buildings, styles, etc., are linked to the becoming random of the referent. Everybody takes the signs of the other to refer to the self. The self is the product of a creation, an artificial construction that mixes and combines signs from the kaleidoscope of the imagination. The self is simulation, becoming other.[4]

The establishment of a capitalist cultural space is marked by a disruption of codes of belonging. Who are you? Where do you come from? Whose son are you? What is your role? What is your value?

Before modernity, in the age of the aristocratic order, social hierarchy was eternally fixed by theological knowledge. Every question had an answer; every person had a place and a value. But everything was displaced when the Copernican revolution asserted that man was no longer the center of the universe, when a new world was discovered beyond the ocean, and when the proliferation of printed texts and painted perspectival images replaced divine creation with human simulation.

Capitalism found its anthropological dimension and its epistemic foundation in this space of groundlessness, which was also a space of freedom. God no longer defined value, and value was no longer something fixed, referring to a natural criterion. Value was the product of labor, time, and power: value was simulation.

Picaresque novels such as *El Buscón* and *Lazarillo de Tormes*, which became incredibly popular during the golden centuries of the Spanish Empire,[5] narrate this disorientation and the effects of this social and cultural deterritorialization. The word *buscón* means *seeker* in Spanish. Lazarillo, the *buscón* par excellence, is someone who is looking for something. But for what? He is looking for food, for money, but mainly for his own identity. The *picaro* represents the passage from a society based on blood and traditional affiliation to a society that is based on money. Blood and tradition were a guarantee of descent and affiliation; they were proof of the link to origin.

Traditional affiliation implied a relationship to the truth (where truth is the origin, or God), a reference to an ultimate reality beyond appearances and beyond mundane fabrications.

Money is the opposite of affiliation and foundation. Money is groundless, and the wealth of the bourgeoisie implied a perpetual shift from one appearance to another. Money is an abstract equivalent, not the signifier of a fixed referent. The *picaro* is nobody's son, in contradistinction to the *hidalgo*, who is *hijo de algo*, the son of someone (from the Latin, *aliquis*). The *picaro* is thus looking for his own identity, and in this search, playing the game of appearances and of simulation.

The baroque search for meaning throughout the vertiginous kingdom of proliferating appearances was doomed to fall into disillusion (*desengaño*) because the general condition of modernity was based on an acknowledgment of the absence of ontological foundations, on the perpetual sliding and shifting from one level of simulation to the next.

When visual space was invaded by ostensible representations, simulated spaces, and perceptual tricks, the infosphere became increasingly thick and complex. Maravall speaks of baroque cosmovision, a cosmology that is based on the relativism of the vision.

The Renaissance [...] saw in the phenomenal world a manifestation or reflection of an objective reality, [whereas the baroque] saw experience as the translation of interior vision. [...] Appearance and manner are not falsehoods but something that in some way belongs to things. Appearance and manner are the face of a world that for us, in any case, represents a phenomenal world [...].[6]

According to Panofsky, perspective was a symptom of the exhaustion of ancient theocracy and of an emergent anthropocentric vision. The possibility of simulating space led to the dissemination of perspectival viewpoints.

The baroque mind forgot that faith was the foundation of imagination, and acquired a new consciousness of the surface, and of the powers of persuasion of artificial simulation.

But there was a political problem in this transition. Creating consensus and propagating faith were goals of the Catholic Church and of the *reyes Catolicos* in the fight against the Protestant Schism, and also underlined the project to evangelize and colonize the new world. In the pursuit of this project, the visual scene of the Counter-Reformation was not based on ascetic contemplation, but on a spectacular profusion of images. Rather than the silence of monastic ecstasy, the condition of *Propaganda Fide* was the noisy explosion of an endless production of images.

Capitalist frenzy started in similar cultural conditions. Since perceptual experience was misleading, and proliferating simulation (of labor, of creation, of invention, of knowledge, of art, and of technologies of reproduction) inundated the mind with semiotic inflation, economics took the place of theology as the ultimate source of truth.

From the Spanish colonization to Hollywood's colonization of the planetary mind, the condition for the capitalist penetration of the collective unconscious has been to saturate the space of the imagination.

Don Quixote wanders within this syndrome of pathological deception. Absorbed by his world of chivalrous fantasies, he is fascinated by its spell and eventually finds himself in a world that no longer exists.

Truth and Measure

While the baroque mind was haunted by the perception of the open interminability of God's creation, and by the ensuing phantasmagoria or plurality of possible worlds, the Protestant imagination—suspicious of the tricky language of images—essentially trusted in the severity of verbal semiosis. The Gothic vision of the Protestant bourgeoisie relied on the perfection of a technological God who spoke an un-ambiguous language: the language of deeds, of mechanical precision, of equivalence, and of measurement. We could also call it the language of reason.

While the baroque drew imaginative energy from deterritorialization (conquest, the proliferation of images, the triumph of the dissipative energy of the imagination), the puritanical core of Protestant culture affirmed that individual consciousness was the site of the uniqueness of truth.

The purity of Puritans was essentially an effect of the erasure of the past, and of historical and religious memory. This erasure created the smooth space of a future without limits that had to be written in a purely technical code.

Modern capitalism was predominantly based on the affirmation of a measurable relation between labor and value. The dynamics of accumulation defined the measure of the value of a good on the basis of the time that was necessary for the production of that good.

The bourgeoisie was essentially a territorialized class. The very definition of this class was related to the territory of the *bourg*, the city, the place where productive energies were assembled, where factories were built and property was protected. Protestant ethics supported a sense of belonging to the local community since this community was witness to the grace of God, attested to by the accumulation of wealth.

In addition, the wealth of the bourgeoisie was territorialized, and the accumulation of capital was enabled by the production of things made of physical materials linked to place, land, and territory.

Working time and territory were the conditions of universal rational measurement. While baroque culture emphasized the ambiguity and the multifaceted, deceptive nature of language, Protestant culture was based on the assumption of a fixed relationship between sign and meaning, signifier and signified.

Puritanism and Virtualization: Neo-Human Culture

Look at the difference between the Catholic style of cultural colonization in Mexico and South America, and the Puritan style of colonization in North America.

After the discovery of the New World, at the beginning of the Spanish colonization of Central America, Bartolomé de las Casas, a religious man who accompanied the Spaniards in their conquest, asked the following: "Should we consider the natives we have discovered in the new territories humans? Do they have a soul as we do? Can they receive the Christian message that can only be transmitted to human beings?"

The Catholic Church's answer was to evangelize the *Indians*, to conquer the soul of the natives, and to allow a certain degree of tolerance for syncretic religious imagination. Evangelization went along with colonization.

The Puritans who occupied North American lands were less tolerant, and their religious activity was less inclusive. They did not ask themselves the baroque question about the soul of the natives. Indians were only seen as an obstacle to expansion, not as people to convert. The people who lived in the northern prairies had to be removed, cancelled, and exterminated. Their obliteration was the condition for the establishment of a new, perfect humanity, whose goal was the realization of the word of God on the smooth space of America. The genocide of the Northern Indians was not an accident, but the distinctive character of an emerging neo-human culture.

The cultural environment in which monetary abstraction could thrive was not Mediterranean territory, densely fertilized by religion and history, but a newly discovered space, inhabited by synthetic faiths and artificial constructions that presupposed the cancellation of historic legacies.

A purified form of economic reason paved the way for the long process of the mathematical mutation of language, which led to the creation of calculation machines, to the mathematical reformatting of the social mind, and, in late modernity, to the shift from an alphabetical to a digital language and techno-scape.

The historically and ethnically purified territories of North America provided the perfect space for the disembodiment of language, and for the process of abstraction that led to digital culture and to virtualization.

The political entity that was generated in the virgin territories of the American utopia, a smooth space oblivious to the roughness of historical and cultural legacies, can be viewed as a neo-human civilization. The continuity of human history was broken by Puritan purification, and humanism was reinstated on the basis of an a-historical truth of the text.

In Puritanical space, language was conceived in terms of pure information: operational truth and the reduction of signifiers to unambiguous signifieds.

In the Puritan view, America was a land without history, and American history was based on the endless motion of the frontier. There was no feudal past or tradition holding back and influencing the creation of a pure community.

The past was nothing, and the frontier everything: a future without contamination or resistance. American Puritanism was first of all the purity of the horizon of time.

> The Unites States is unique in the extent to which the individual
> has been given an open field unchecked by restraints of an old
> social order, or of scientific administration of government.[7]

The mythology of roots was cancelled, which had been so important in Euro-Asian history, and the new world was conceived as a place where choices were perfectly binary: good or evil, light or darkness.

Purified of the emotional reference to the body and to ethnic memory, civil and religious spheres had to merge so that perfect reason could reign.

> Always the mere emotion of religion was to be controlled by
> reason. Because of this, the university-trained Puritan clergy prided
> themselves on the lucidity and rationality of their sermons.[8]

The Puritan spirit, diluted in the denominational field of American Protestantism, deeply marked both religious and civil culture since the American national culture did not recognize the distinction between religious and civil structures.

Technology was religiously conceived as *eupraxia*, the correspondence of human and divine art. A mentalist and functionalist God could help humanity, provided that it had purified its mind from any residual cultural past. Purification and emancipation from history paved the way to an epistemological space where the abstraction of code could be conceived and the operational potency of technology deployed.

The obsession with purity translated reality into written code. Only what could be verbalized was real, and all of reality could be verbalized. What could not be verbalized was demonic suggestion. God pronounced words, and at the beginning was the *verbum*, a purely operational language.

> The primary problem of Protestantism is word-fixation: Scripture-study is at its heart. No fleshy mediator is needed between the soul and God; no images of saints, Mary, or God are permitted. [...] In highly ritualized Italian and Spanish Catholicism, by contrast, there is a constant, direct appeal to the senses.[9]

The purification of language from bodily ambiguity was a pre-condition for the reduction of social relations to calculation, and for the superior potency of capitalism in the field of operational efficiency.

All the trans-historical *enterprise* had to face was an endlessly moving frontier, without contamination or resistance. American Puritanism implied the purity of the horizon first of all. The universe was smooth and prepared for a binary, ethical perception. The world was the emanation of a verbal flow, and God was the original code that could act on the minds of men, provided that they were pure, clear from the impurities of cultural becoming.

The Manifest Destiny of the neo-human entity that is embodied in the United States of America is this: to cleanse the world, to remove human imperfection, and to perfect the identification of the world with God's will, a will that is written in plain, unambiguous words.

Resurfacing the Baroque in Post-Bourgeois Semiocapitalism

Indust-reality and the mechanization of labor shaped the perception of the modern world, culminating in the aesthetics of the modernist avant-garde.

The coincidence of the avant-garde with industry and the media led to the proliferation of the art-object, the overlapping of art and production, and thus to the aestheticization of everyday life, all distinctive features of late modernity. Aesthetics invaded mass production, and every real object, losing its singularity, appeared as the reproduction of a model. From this point of view, late modernism could be viewed as the process of replacing an original reality with artifacts. Art was dead, said Jean Baudrillard, not only because critical transcendence was dead, but also because reality itself was being confused with its image.

Since then, the process of digitalization has transformed things into signs, and objects into messages. The proliferation of semiotic goods has produced a saturation of social attention, a neo-baroque effect. The passage from modern capitalism to semiocapitalism is marked by the end of measure, and by the return of the spirit of the baroque.

In his book *Vuelta de siglo*, Bolívar Echeverría argues that the spiritual and immaterial power of the Church of Rome had always been based on the ideological control of the imagination, but this influence was hardly considered by the pragmatic ethics of industrial culture.

Catholic Spain of the golden centuries was the harbinger of a non-industrial brand of accumulation based on the massive pillage of the Americas. This strain of modernity was marginalized after the military defeat of the Invincible Armada in the naval war with the British Empire, which started Spain's economic and political decline. Protestant modernity defined the canon, but the baroque strain of modernity was not erased. It went underground, tunneling deeply into the recesses of the modern imaginary, only to resurface at the end of the twentieth century, when the capitalist system underwent a dramatic paradigm shift towards post-industrial production.

Semiocapital is centered on the creation and commodification of techno-linguistic devices that, by their very nature, are semiotic and deterritorialized.

We must grasp the social implications of the two different strains of modernity. The relationship between the industrial bourgeoisie and the working class was based on conflict, but also on mutual cooperation. Despite the radical conflict opposing salary and profit, the bourgeoisie and the working class could not dissociate their destiny, because they lived in the same physical space, and needed each other in order to survive and to grow. Although the alliance between labor and capital broke down in the 1960s and in the 1970s, a new alliance was possible in the last decade of the twentieth century. The experience of the dotcom enterprises was the expression of an alliance of venture capital with intellectual work, and this temporary convergence of engineers, artists, and venture capitalists made the extraordinary technological progress of the digital sphere and the creation of the Internet possible. But this alliance was broken when financial power prevailed on cognitive labor, and when the financial predatory machine invaded the empty space of randomized value.

When language becomes the general field of production, when the mathematical relation between working time and value is broken, when deregulation destroys all liabilities, predatory behavior becomes the norm. This is what has occurred since financial capitalism has occupied the world scene. Deregulation must be seen in the context of the technological and cultural evolution that displaced the process of value creation from the field of mechanical industry to that of semiotic production. Since cognitive labor is hardly reducible to the measure of time, the relation between working time and valorization becomes uncertain, indeterminable. And when the relation between labor and value can no longer be decided, what reigns in the global labor market is the pure law of violence and abuse.

Neo-Baroque and Deregulation

Simulation and fractalization—features of semiocapitalism—are essentially baroque categories. In his book *Neo-Baroque: A Sign of the Times*, Omar Calabrese claims that postmodernism recuperated aesthetic and discursive models that first emerged in the 1600s. The baroque is essentially the proliferation of points of view, and the contemporary digitalization of labor has shifted production and exchange into a regime of indetermination.

Since Nixon's 1971 decision to deregulate the American dollar, emancipating it from the gold standard, the American economy has been freed from the rule of law, and American debt has been allowed to grow indefinitely, since the debtor had stronger military power than the creditor. Far from being the subject of an objective science, in this case, economics shows itself to be an enterprise of violent obligation that models social relations. In *Symbolic Exchange and Death*, Baudrillard writes the following:

> The reality principle corresponded to a certain stage of the law of value. Today the whole system is swamped by indeterminacy, and every reality is absorbed by the hyperreality of the code and simulation. The principle of simulation governs us now, rather than the outdated reality principle. We feed on those forms whose finalities have disappeared. No more ideology, only simulacra.[10]

The entire system has precipitated into indeterminacy, since the correspondences between referent and sign, simulation and event, value and working time are no longer guaranteed.

Since the 1970s, the shift to immaterial production has been eroding the identification of wealth with physical property, and the identification of value with territorialized labor.

Deregulated semiocapital has marginalized the bourgeoisie, replacing the latter with two distinct and opposing classes: the cognitariat, i.e., the precarious and cellularized cognitive workers, and the managerial financial class, whose only competence is competitiveness and financial manipulation.

The process of deregulation has paved the way to the baroque space of randomness. Paradoxically, the baroque imaginary can only be productive when the Puritan spirit has utterly reduced language to algorithmic operational processes. When the purification of capital accumulation has been accomplished by finance, the baroque enters the scene of hyper-real production.

Iconophobia

Pagan cultures often attribute to the image a value of trans-reality. Images may have magic effects. Pagan gods are not jealous of images because their divinity resides in the image itself. The image is part of the animist continuity that pervades both nature and artifact. The spirit cannot be separated from its environment.

Iconophobic sentiment has two main sources. The first is Platonic philosophy, which implies a contradiction between an image and a truthful idea; the second is the biblical prohibition against the replication of images.

In monotheism, the word (*verbum*), the voice, and the Law are the semiotic forces that define religious psycho-culture.

"Listen, Israel," says the God whose name cannot be pronounced. The word is His force, and by the word he creates and governs the world. God spreads terror through the word; through the word, he consoles. And in fact His name cannot be contained in a word.

Jewish iconophobia was accompanied by an obsessional deification of the word. In the Jewish diaspora, the literary tradition played an important role, making the transmission of a strong cultural identity possible. This explains why Jewish visual art does not really exist, although Jewish artists do exist.

Jewish painters such as Chagall have taken their inspiration from the Christian mythological and iconographic repertoire, while other painters, such as Mark Rothko, have chosen the way of abstraction and of non-figurative expressionism. Jewish iconography is a kind of non-existing space, filled with calligraphy, words, and inscriptions.

The scriptures are explicit on this point:

I am the Lord your God, who brought you out of Egypt, out of the land of slavery.

You shall have no other gods before me.

You shall not make for yourself an image in the form of anything in heaven above or on the earth beneath or in the waters below. You shall not bow down to them or worship them; for I, the Lord your God, am a jealous God, punishing the children for the sin of the parents to the third and fourth generation of those who hate me, but showing love to a thousand generations of those who love me and keep my commandments.[11]

Nothing among those things that exist in nature or in heaven can be represented by the hands of man. But a question remains. Is one allowed to represent what does not exist in nature or in high heaven, but only in the human imagination? Can the unreal be represented? What about dragons and chimerical beings, what about the phantasmagoric imagination? And what about decoration, the visual play of abstract geometries? Can such inexistent things be painted? From this viewpoint, Chagall could be considered a Jewish painter who does not represent anything that really exists, but only paints dreams, hallucinations, and fantasies. Both geometrical abstraction (Mondrian) and expressionist abstraction (Chagall) are certainly influenced by the need for painting what exists only in the human mind.

The Veil and the Face

In Islam, the prohibition against visual representation only becomes explicit in the third generation after the Prophet. But according to André Grabar, author of *L'iconoclasme byzantin: dossier archéologique,* the horror of images has been a constant feature of Muslim civilization since the beginning.

An an-iconic purism seems to be inscribed in the cultural code of Islam, something reminiscent of the desert, the un-representable space of emptiness. Only the perfect harmony of geometric forms emerging from mental abstraction deserves to be painted and seen, and not the irregular, changing forms of the human body and of physical matter. Abstract decoration and calligraphy are permitted forms of vision in this cultural space.

> A Sufi master declared: "I have never seen anything without seeing God in it." And another: "I have never seen anything but God." It is as if the invisibility of God resulted in the invisibility of things made to be seen [...][12]

The Muslim an-iconicity seems to be rooted in the absolute transcendence of divinity, and does not come as an effect of the prohibition and repression of the imaginative faculty. A wide range of profane iconic production is therefore tolerated in Islam.

Decorative calligraphy appears as the highest form of artistic symbolism. Beauty inheres in the calligraphic gesture. In the sixteenth century, Qadi Ahmad writes, in the *Gulistan-i-hunar (Rose Garden of Art)*: "The essence of writing is in the spirit. [...] Excellent writing clears the eyes [...] writing is the geometry of the soul."[13]

As writing creates the world as world of the soul, writing is the geometry of the world. Creation itself can be conceived as a calligraphic gesture that decorates the pages of time according to God's design. God is the first painter, or rather the first calligrapher, as creation seems to be more an intellectual act of conception and writing than emanation or the projection of images.

Rumi, the most celebrated Islamic poet, writes that God is the calligrapher who draws his works with the pen of the human heart. In the book *Soufisme et art visuel*, Laibi Sakir outlines the psychological dynamics implied by the Islamic rejection of figuration, and by the marginal, almost secretive flourishing of talismans, objects with strong magical undertones.

> Islamic art [...] settles on the exclusion of what it considers secondary, beginning with the human form, and through the rejection of all representation of objective reality, in favor of what it considered the only true reality: that of the invisible.
>
> The great paradox of Islamic art resides in its desire to represent a reality situated outside of the visible real [....].[14]

The talisman, which already existed among the Chaldeans, the Egyptians and the Latins, suggests the surfacing of a representative visuality that is not allowed in the recognized sphere of art. What is denied in the sphere of official art—the visual representation of the body—reappears as a transgressive experience in Sufi mystical experience. The body (*jism*), shape (*surah*), and the visible (*ruy'a*) are linked by a relation of concealment, denial, and transgression. The Islamic aesthetic sensibility is grounded in this dialectic tension, which can be expressed by the veil as a form of concealment and a simultaneous hinting toward the concealed visible. According to Dominique Clévenot:

> The face of God is hidden, although believers turn their gaze towards it. The Qur'an defines believers as those who desire to see the face of God. Here is the core of the Muslim conception of the veil, a screen that encloses vision, but simultaneously exalts desire.[15]

Clévenot shows that the Islamic psychosphere is based on the shadowing of the visible. Traditional urban landscape can also be described from this point of view:

> In these cities, the street is a series of visual breaks. [...] The traditional urban house of the Muslim-Arab world [...] tries to escape the gaze. Entirely turned inward to its interior courtyard, [...] on the outside, it only reveals a few sections of wall.[16]

The veil, an obstacle placed between the face and vision, is a crucial component of the Islamic imaginary. Visibility can cause danger, vertigo of the mind, and fatal attraction.

> God has seventy (or seventy thousand) veils of light and darkness. Without these veils, the radiance of His face would no doubt consume everyone who saw it.[17]

Hagar

Investigating the problem of the veil in the context of a wider consideration of femininity in the Islamic world, Fethi Benslama evokes a denial (*refoulement*) of the visibility of women, and more generally, of the singularity of the body.

The mobility and visibility offered to women by the global media have exacerbated the reactive attitude of men in the Islamic world.

In *Psychoanalysis and the Challenge of Islam*, Fethi Benslama speaks of the nostalgic torment of the origin as a peculiar feature of Islamic psycho-culture. According to him, the *torment* of the origin manifests in Islam in the form of the suppression of the feminine, and particularly in the denegation of the person of Hagar, who, in spite of her foundational mothering role, is never mentioned in the Qur'an, and is also *evicted* from the Biblical narrative so that Abraham the father can be reconciled with his son. Hagar was a slave that gave Abraham a son named Ismail when Sarah, his legitimate wife, was considered infertile. Later on, according to the Bible, Sarah became pregnant and gave birth to Isaac. After Isaac's birth, Hagar, who had never been fully accepted into the tribe despite being the mother of Abraham's child, was completely rejected, and driven out into the desert. Hagar means *stranger*. She was from Egypt, and when she became pregnant, God promised that her child would be the ancestor of a great nation. This nation is the Arab people.

Benslama underlines the fact that the theological foundations of Islam are marked by the identification of women as danger. The Qur'an (19:28) speaks of the immense trickery of women, who have neither reason nor religion. The visibility of a woman's face is the origin of guilt, because the male is unable to control his own reaction to vision.

According to Benslama, the Islamic rejection of visibility, and particularly the visibility of women's faces, is not, as in Platonic philosophy, a consequence of the deceptive feature of the visible. On the contrary, vision and truth have a direct relation: seeing opens the eyes to the vision of truth, and truth has to be hidden.

> Veiled, unveiled, reveiled, these are the three stages in the female operation of theology: initially veiled, unveiled to demonstrate originary truth, then reveiled by order of the belief in that truth of origin. For, once established, truth aspires to conceal the nothingness through which it has passed.[18]

Women's visibility is a danger for masculine identity and for the community itself.

> Like an uncontrollable visual orifice, he can be penetrated by female monstrations, which possess and subjugate him to the extent that he forgets his law. [...] Theological discourse has continued to present us with man's confusion around woman, a disturbance that is related to woman's clairvoyant power, her knowledge of alterity, and the originary confusion in the face of the identity of her being.[19]

Hiding the disturbing truth is the true motivation of the Muslim veiling that permeates daily life and political relations. The truths of the body, of singularity, and of desire have to be denied and hidden in order to protect the cohesion of the community. Muslim culture is essentially based on the assumption that the individual draws legitimacy and psychological consistency only by belonging to the community. Marks of singularity are therefore to be hidden and possibly cancelled.

Truth, Effectiveness, Iconocracy

In the Christian world, both orthodox iconophilia and iconoclasm share the same furious passion for transcendence. Humanist representation emerged from the abandonment of that passion, and from a richer and more multifarious interest in the reality of human life. According to Mondzain, in *Image, Icon, Economy*, the icon should not be mistaken for a representation. In the icon, what counts is not the object of representation, but its lack, spiritual absence.[20] The icon is the actual figuration of the impossibility of figuration. Mondzain retraces the history of the general system of visibility starting from the war between iconoclasm and iconophilia in the Byzantine age.

The relation between the visible and the invisible is at the heart of the history of the economy, according to Mondzain. In fact, the economy is the sphere in which material things that are visible, touchable, and usable are interpreted and exchanged in terms of abstraction, that is money and value.

Orthodox sensibility is essentially focused on truth, interiority, intensity, and purity—while the Catholic version of religion aims at the effectiveness of the word, of the image, and of gesture.

The Eastern Christian world is divided between iconoclasm and iconodulism in the name of truth. For the iconoclasts, the image has to be rejected because it is a lie; for the iconodules, it has to be embraced as a step on the path of ascetic experience, as the sign of an absence, and also as a crutch for the majority of the faithful.

But in the period that followed the Protestant Schism, the Catholic Church no longer thought in terms of truth, but in terms of persuasion and effectiveness. The baroque innovation consisted essentially in that. The image was an artifact, and an artifact could not be judged upon the presupposition of truth or falsity. Rather, it had to be judged in terms of effectiveness.

Effectiveness is not limited to the problem of the propagation of faith, but refers to the creation of power, political consensus.

Here we find the foundation of modern iconocracy, the power of images.

In the modern age, the image invaded urban space thanks to technologies of reproduction, thanks to printing machines and electricity, and at the end, thanks to digital simulation. Economy and the spectacle developed together.

The globalization of images runs parallel to economic globalization, and the integration of the image in the process of valorization is simultaneous to the subsumption of human creativity and intellectual potency by the general form of semiocapital. We can thus view modernity as an iconocratic age.

Visual Culture and Globalization

Over a short span of time, being immersed in a video-connective environment, the human mind has acquired cognitive competences such as the capacity to read electronic images, to understand the meaning of recombinant flows of visual information, and to interact with the connective machine. Modes of cultural and psychological elaboration, however, have not adapted as fast.

Identitarian reterritorialization is often a response to techno-imaginary deterritorialization. This reaction led to the surfacing of manic-depressive pathologies such as panic, depression, and the political pathology of massive fanaticism. This is why the vision of a peaceful global Empire, which prevailed in the 1990s, failed. Shared by techno-liberal thinkers such as Kevin Kelly and radical Marxists such as Toni Negri, this vision was based on the assumption that democracy and technological implementation of the intellectual potency of the social brain were within range.

After the dotcom crash of April 2000, and after the World Trade Center bombing of September 2001, war took over the center of the worldscape; although globalization had not receded, it had already re-codified both the imaginary and the techno-cognitive styles of communication. The deterritorialized Empire provoked a resurgence of aggressive territoriality, but territorial identities could only express themselves in the deterritorialized language and through the deterritorializing technical media of the global Net. And this language is mostly visual.[21]

According to McLuhan, the transition from the modern industrial environment to the electronic one implies that *configuration* prevails within culture. As the sequential pattern of printed writing is replaced by the configuration of images, the social mind tends to replace the critical with a mythological methodology of interpretation. Critical evaluation is possible when an exchange of signs is slow enough to be sequentially scrutinized by a reader, and when a receiver has time for the discrimination between true and false that is called critique.

Visual information, particularly the electronic flow of configurations, is too fast for critical examination. The visual mind thus tends to understand this flow according to mythological frameworks of interpretation.

The visual—in this context—has little to do with representation. The concept of visual culture is not about representational forms. It is about the mode and speed of emanation and reception of any kind of sign. Words and texts become part of visual culture when their exposure is accelerated so that they are synthetically apprehended as visual stimulation, rather than sequentially decoded and interpreted.

Written text is not disappearing under the effect of electronic media technology. Cellular text messaging, indeed, has enormously increased the amount of written text circulating in the infosphere. But text messaging becomes part of the visual, and demands visual configurational procedures of reception and interpretation. Writing techniques change, and become increasingly synthetic, configurational, fast, uncritical, and mythological. Think of Twitter.

Images work as activators of cognitive chains that transcend both the limits of verbal language, and local national interpretative grids. The more a cultural content can be translated into images, the more it can be widely disseminated, as images are less dependent than words on the territoriality of a national language.

Visual configuration has fostered the integration between Western culture and a Far Eastern culture based on ideographic writing. Advertising, television, Hollywood and Bollywood block-busters, rather than books, have been the bearers of globalization.

The process of globalization occurs at a much deeper level than the one on which political processes develop. The collapse of the Soviet Empire was not an effect of political persuasion, but rather the result of the cultural globalization that electronic media made possible in the 1980s.

In the book *Images at War*, the anthropologist Serge Gruzinski retraces the origins of the process of visual globalization starting from the meeting of pre-Colombian Mexican visuality with the baroque.[22] The visual legacy of indigenous populations, linked to the use of hallucinogenic substances and to a multifarious religious imagination, was overpowered and absorbed by Christian mythology and imagination. However, this did not cancel indigenous visual and mythological memory, but instead provoked a contamination that led to a reframing of the Catholic doctrine itself.

The suggestive power of the baroque imaginary penetrated the imagination of local populations. Indigenous people reacted to the Catholic repression of pagan images with manipulations, translations, and assimilations that resulted in the rich imaginary syncretism that is the popular culture of a large part of Latin America.

In the imaginary, the principle of contradiction does not operate, and incompossible prospects can coexist, as they do in the unconscious according to Freud.

PART 2

THE BODY OF THE GENERAL INTELLECT

What is the history of the relation between techno-scientific potency and the social body, and in particular, what is the genealogy of financial capitalism from the point of view of language and of the technical evolution of production? According to Marx, the expression *general intellect* refers to the development of scientific and technical knowledge as a force of production.

> The development of fixed capital indicates to what degree general social knowledge has become a direct force of production, and to what degree, hence, the conditions of the process of social life itself have come under the control of the general intellect and have been transformed in accordance with it.[1]

As the general intellect develops as a productive force, it is captured by the abstract machine of semio-valorization, and is separated from its social and affective body. The capture of the general intellect, and the submission of knowledge to the rationale of the profit economy is the defining feature of semiocapitalism.

Cognitive workers—the cognitariat—suffer from a new kind of alienation, issuing from the separation of virtual activity from bodily existence and communication.

The recomposition of the general intellect as a social and political force is only possible when the intellectual potency of labor rejoins with the affective body of the cognitarians.

The process of recomposition of cognitive work therefore simultaneously passes through language and the body.

4

Language, Limit, Excess

Prophecy, Imminence, Possibility

If we try to map a phenomenology of the contemporary aesthetic-scape, we notice a prevailing dark prophetic imagination. Prophecy is not an abstract prediction of the future; rather it is about imminence.

Imminence is what the present reveals about the future, the horizon where the present discloses a looming possibility. The imminent is immanent, inscribed in the present. There is no transcendence involved in imminence. Every possible evolution is already here. But the possibilities are many, and different.

Immanence is the quality of being inside the process. It is the inherence of one thing to another. The future is inscribed in the present structure of the world as a range of possibilities, as a tendency that we can imagine, sense, and perceive, although we cannot clearly see it. Prophecy indeed is a sort of premonition, a vibratory sense of intuition that gives us the possibility of feeling the future as tendency. The present state of the world can be described as the vibratory concurrence of many possibilities. How will chaotic vibration give birth to the event? Why is it that among many possibilities of development one of them prevails?

The relation between now and tomorrow, between the present and future states of the world is not one of necessity. The present does not contain the future as a linear development. The emergence of one form among many possible forms is the provisional, unstable effect of a polarization, the fixation of a pattern. In the social field, the emergence of one form among many possible forms is the effect of a reduction in the randomness of subjectivity.

Aesthetic sensibility is about tuning into the vibratory plurality of possibilities, and detecting a possible inclination in terms of their evolution. The poetic act is the prophetic hint of the *clinamen*, of the point of precipitation and fixation of the vibratory movement of possible events.

In contemporary aesthetic production, the signs of a dark *zeitgeist* are easy to detect.

The *zeitgeist*, the spirit of the time, involves a perception of imminence.

A dead-end imagination is widely at play in the aesthetic phenomenology of our time. Art, poetry, narration, music, and cinema trace a landscape of imminent darkness: social de-evolution, physical decay, and neuro-totalitarianism.

Every prophecy is in some degree a self-fulfilling prophecy, because prophecies predispose the social imagination to expect a certain evolution, and shape common perception such that what is largely foreseen becomes common sense.

The art-scape of the new century seems crowded with dystopian imaginings, depressing descriptions of the present, and frightening scenarios of the imminent time to come.

I am referring here to blockbuster movies such as *The Hunger Games*, but also to the work of sophisticated filmmakers such as the Korean Kim Ki-duk (*3-Iron, Pietà*) and the Chinese Jia Zhangke (*Still*

Life, A Touch of Sin), as well as novels such as *The Circle* by Dave Eggers or *The Corrections* by Jonathan Franzen. A cynical consciousness seems to blow from the contemporary imagination, signaling the resigned acceptance of the unavoidable misery of precarious life.

Nonetheless, the poetic act may also be the aesthetic experimentation of a shift in the semiotization of the world. It can transform the sensitive algorithm that perceives and projects the world. Narration and poetry exhibit forms of action and adventure that can be reenacted in life. They are mythopoiesis, the creation of mythologies to reshape expectations.

Prophecy is the enunciation of a perceived tendency, a tendency being the movement of the environment towards a given direction. The vibratory complexity of the world can be interpreted in terms of coexisting, conflicting, and interacting possibilities. A tendency is the possibility that seems to prevail at a given moment of the vibratory process, leading to the event, to the next possible geography.

The poetic act may be conceived as a dispeller of gloom, as a semiotic engine aimed at dispelling the self-fulfilling prophecy of depression.

Poetry as Excess

Why do human beings address words, sound, and visual signs poetically? Why do they shift away from the level of semiosis? Why do they subtract signs from the established framework of exchange?

Hölderlin answers this question in his late poem "In Lovely Blueness,"

> Full of merit, yet poetically, man
> Dwells on this earth [...].[1]

Heidegger and other philosophers have discussed Hölderlin's text in their way. My intention here is to find a dimension of the poetic act that may be seen, in Hölderlin's words, as opposed to the *merit* of man.

What is merit? I would say that merit is the quality of being worthy, of deserving praise or reward, the quality of satisfying the measure (the conventional measure) that concerns the (conventional) value of individuals interacting with each other on the social scene.

Social beings are more or less full of merit. They deserve recognition as they exchange words and actions in a meritorious way, and receive a kind of moral payment that is mutual understanding and the confirmation of their place in the theatre of social exchange.

But merit, moral payment, and recognition belong to the conventional sphere. When humans exchange words in social space, they presume that their words have an established meaning, and produce a predictable effect. However, we can do things with words that break the established relation between signifier and signified, and open new possibilities of interpretation, new horizons of meaning.

In the last lines of Hölderlin's poem, he writes:

> Is there a measure on earth? There is
> None.

Measure is only a convention, an intersubjective agreement that is the condition of merit, or social recognition. Poetry is the excess that breaks the limit and escapes measure.

The ambiguity of poetic language is indeed an effect of semantic over-inclusiveness. Like the schizo, the poet does not respect conventional limits in the relation between the signifier and the signified, and reveals the interminability of the process of attributing meaning.

Excessiveness is the condition of this revelation.

The world is the effect of a process of semiotic organization of pre-linguistic matter. Language organizes time, space, and matter in such a way that they become recognizable for human awareness. This process of semiotic emanation is not a natural given. Rather, it is a perpetual reshuffling, a continuous reframing of the environment. Poetry can be defined as experimenting reshuffling and matching patterns of emanation with projections of the world.

The act of defining that I have just performed is arbitrary and illicit, because the question "What is poetry?" cannot be answered. I cannot say what poetry *is*, because in fact, poetry *is* nothing. We can only try to say what poetry *does*.

The act of composing signs, be they visual, linguistic, musical, etc., may disclose a space of meaning that does not exist in nature, and that is not based on a social convention.

The poetic act is the emanation of a semiotic flow that sheds a light of non-conventional meaning on the existing world. The poetic act is semiotic excess hinting beyond the limits of conventional meaning. Simultaneously, it reveals a possible sphere of experience that was not previously experienced: the experienceable. It acts on the limit between the conscious and the unconscious in such a way that this limit is displaced, and that parts of the unconscious landscape—the *inner ausland*, the intimate foreign country—are lit up, or distorted, and re-signified.

That said, I have said nothing, or almost nothing, or very little. Poetry is the act of language that cannot be defined, since de-fining means putting limits. Poetry is precisely the excessiveness beyond the limits of language, which are the limits of the world. We cannot define poetry, or poetic acts. Only a phenomenology of poetic events can give us a map of poetic possibilities.

Word and Mantra

Semiotic production occurs between the poles of representation and immersion. The sign can act as a re-presentation of reality, or as an activator of associative chains and a stimulator of sensibility. In the first case, the artist is active, and those who surround him are only spectators. In the second case, art is the nootropic source of mental stimulation, directly acting on perception, cognition, and behavior. The artist acts like the shaman who knows the rules of the ritual, and the effects of the substance (both organic and semiotic) that is able to trigger a psychedelic process of mental stimulation.

Being immersive is an intrinsic feature of music. Music does not *signify*. Instead, it acts to stimulate and immediately modify states of mind without the mediation of signification. It was for this reason that symbolism emphasized the immersive effect of words, their sonic matter, and their audible sensuousness. Musicality is most important in poetry because the evoking potencies of musical vibrations directly act upon the mind. In musical action, signifier and signified are not separated, and music triggers effects of a-signifying immersive stimulation.

> Poetry is subtler than prose, because its rhythm produces a higher unity and loosens the fetters of our mind. But music is subtler than poetry, because it carries us beyond the meaning of words into a state of intuitive receptivity.[2]

Alexander Blok, the symbolist poet who was so important in the Russian literary landscape in the first decade of the past century, writes:

Music is the most perfect of the arts, primarily as it expresses and reflects the design of the Great Architect to the fullest extent [...] Music creates the world, and is the spiritual body of the world, fluid thought [...] poetry, reaching beyond its limit, will probably drown in music.[3]

Music is the ritualization of noise. It is ritual without a meaning. René Daumal, speaking of Indian music in the book *Bharatha*, starts from the distinction between *mot* (word) and *parole* (speech). "La parole est plus réelle que les mots." (*Speech is more real than words.*)

Parole in this context is the non-denotative word, sensible stimulation, which differs from the word as denotation, *mot*. The difference between these two French terms, according to Ferdinand de Saussure, is the difference between sonic materiality and denotational signification. In *parole*, the singularity of the voice resounds. Western music has internalized the rule of sonic succession, in the form of melody, in the construction of *contrappunto*, and in conventional notation. This formalization tends to separate music from the singularity of the voice as an event, an unrepeatable situation in time.

Indian music has resolved its relation with melody in a peculiar way. The individuality of Western musicians is manifest in the creation of an original melodic succession, while the Indian musician sticks to the rules and melodic sequences established from time immemorial, as an ancient tradition has limited the range of available musical themes. So, since its melodic structure is minutely detailed, rag music is about the singular vibrations of the interpreter. Interpretation is not a re-actualization of a musical score, as it is for Western musicians, but a singularization of a traditional tune.

Raghava R. Menon writes:

A note-based tradition has no use of personal transmission in the style of the Guru, as there is an external mechanical measure of musical sound whose personal characteristics are minor and merely interesting. The note standardizes music; *Swara* liberates it. It is in this liberation that the secret of the Guru lies, and his vital connection with the preservation of the flaming interior of the art.[4]

The musician's performance is a singular, unrepeatable event. Music is the ritualization of sound and the ritualization of silence, building rhythmic pathways and sonic constellations through the infinite noise of the universe.

Félix Guattari speaks of the ritornello, a singular a-signifying rhythm that helps to determine orientation in the world.[5]

According to the Italian music critic Salvatore Sciarrino, European romanticism transformed music from a time-based machine into a space-based machine. Traditional music flows from beginning to end as a quiet river, while the classic symphony is a spatial construction. Symphonic music obeys structural rules, patterns that contain the natural flow of its river. Beethoven is the watershed between music as flow and music as highly structured form. According to Sciarrino's analysis, music has passed from acoustic and time-prone to visual and spatial.

According to psychoanalysis, writes Mircea Eliade in the book *Images and Symbols*, "the dramas of the modern world proceed from a profound disequilibrium of the psyche, individual as well as collective, brought about largely by a progressive sterilization of the imagination."[6] Eliade argues that linguistic standardization is linked to the reduction of language to a mere tool for exchange. Eliade's words echo Anagarika Govinda's:

> In this age of broadcasting and newspapers, in which the spoken and the written word is multiplied a millionfold and is indiscriminately thrown at the public, its value has reached such a low standard, that it is difficult to give even a faint idea of the reverence with which people of more spiritual times or more religious civilizations approached the word, which to them was the vehicle of a hallowed tradition and the embodiment of spirit.[7]

Semiotic overproduction and the reduction of signs to a mere tool for exchange dull the evocative richness of the word, although, according to Eliade, symbols never disappear from the deep psychological sphere of human beings.

As Viktor Shklovsky wrote in his *Theory of Prose*, literature has the power to reactivate the vital power of words, dulled by daily use, returning them to their original epiphanic potency.

Similarly psychedelic substances have the power to give new life to perceptual, that is, visual and auditory experiences as well as experiences of closeness that have lost their authenticity and freshness because they have been repeated and rendered banal.

Psychedelic drugs allow the mind to evoke visions correlative of perceptual illusions. In fact, the illusion is simply the neural recombination of perceptual engrams, fragments of previous experiences that have been memorized. This recombination takes epiphanic form. Epiphany, from the Greek *epiphainein*, means an apparition, the sensible manifestation of evoked reality. In an epiphany, meaning appears to be perceivable, touchable, visible, and audible, and words are not mere denotative indicators, but activators of a chain of sensorial reactions.

Anagarika Govinda distinguishes between the Indian words *shabda* and *mantra*. *Shabda* refers to ordinary words, used to denote objects and concepts in the normal exchange of operational meanings. *Mantra*, on the contrary, triggers the creation of mental images, and of sensible meanings.

> In the word *mantra*, the root *man* = "to think" (in Greek "menos," Latin "mens") is combined with the element *tra*, which forms tool-words. Thus *mantra* is a "tool for thinking," a "thing which creates a mental picture." With its sound it calls forth its content into a state of immediate reality. *Mantra* is power, not merely speech which the mind can contradict or evade. What the mantra expresses by its sound, exists, comes to pass. Here, if anywhere, words are deeds, acting immediately. It is the peculiarity of the true poet that his word creates actuality, calls forth and unveils something real. His word does not talk—it acts![8]

A *mantra* is a vocal emission that has the power to create mental states with no conventional signification.

Modern poetry, since the end of the nineteenth century, has been a conscious attempt to emancipate the word from its denotational function, aiming to emphasize the epiphanic, immersive value of the word as *mantra*.

Reviving the evocative power of the word, symbolism generated a new sensibility that, paradoxically, anticipated both the technical virtualization of media communication and the reactivation of the bodily singularity of language.

I want to emphasize this symbolist ambivalence. Symbolist poets cancelled the dependence of the word on its referent, while simultaneously celebrating the bodily vibration that arises from the word.

What good is the marvel of transposing a fact of nature into its almost complete and vibratory disappearance with the play of the word, however, unless there comes forth from it, without the bother of a nearby or concrete reminder, the pure notion.

I say: a flower! And outside the oblivion to which my voice relegates any shape, insofar as it is something other than the calyx, there arises musically, as the very idea and delicate, the one absent from every bouquet.[9]

Symbolism opened a new space for poetic praxis. Access to this space occurred by emancipating the word from its referential task.

Poetry is the language that exceeds exchange. It is the infinite return of hermeneutics, and the reactivation of the sensuous body of language.

By poetry, I mean poetry as an excess of language, as a hidden resource that enables us to shift to the suggestive dimension of language.

This is why symbolism acts in an ambivalent way. It triggers emancipation from the referent, and therefore subjects to experimentation the semiotic space to which financial abstraction belongs. However, it also acts as *mantra*, the musical reactivation of the sensorial body that pulses in language.

The symbolist word wanted to be free from the referent, opening the way to a process of evocation, but simultaneously, it wanted to act as bodily vibration, to reactivate the sensuous bond that linked sound and meaning: the voice, the point of conjunction between flesh and meaning.[10]

The Emancipation of the Sign

I want to investigate here the genealogy of semiocapitalism, and particularly of financial capitalism from the point of view of linguistic sensibility and its transformations. Symbolism was an experiment in the emancipation of the poetic sign from the referent, and in the attribution of an evocative power to the sign. The symbolist word was not the representation of an object, but the evocation of an effect of signification.

What interests me is the analogy between the symbolist revolution in the sphere of language and the revolution that Internet finance brought about in the sphere of the economy.

Money and language have something in common: they are nothing and they move everything. They are nothing but symbols, conventions, *flatus vocis*, but they have the power to persuade people to act, to work, and to transform physical things.

> Money makes things happen. It is the source of action in the world and perhaps the only power we invest in.[11]

In Chapter 14 of *Understanding Media*, "Money, The Poor Man's Credit Card," McLuhan writes:

> "Money talks," because money is a metaphor, a transfer, and a bridge. Like words and language, money is a storehouse of communally achieved work, skill, and experience. Money, however, is also a specialist technology like writing; and as writing intensifies the visual aspect of speech and order, and as the clock visually separates time from space, so money separates work from the other social functions. Even today money is a language for

translating the work of the farmer into the work of the barber, doctor, engineer, or plumber. As a vast social metaphor, bridge, or translator, money—like writing—speeds up exchange and tightens the bonds of interdependence in any community.[12]

Marx spoke of money as a general equivalent, the translator of anything into every other thing. This was correct, and still is—but money is not only a signifier whose signified is infinitely various. Money is also an engine, a mobilizer, a source of energy that transcends referentiality and measurability.

Financial capitalism is based on autonomizing the dynamics of money, and on autonomizing the production of value from the physical manipulation of things and the physical interaction of persons.

Here lies the analogy between the history of poetry as semiotic laboratory, and the history of the late modern economy, transiting from industrial capitalism to semiocapital.

Jean Baudrillard proposed a general semiology of simulation based on the premise of the end of referentiality in the economic as well as in the linguistic fields. In *The Mirror of Production*, he writes:

> [...] need, use value, and the referent "do not exist." They are only concepts produced and projected into a generic dimension by the development of the very system of exchange value.[13]

The process of autonomization of money, which is the peculiar feature of financial capitalism, can be inscribed in the general framework of the emancipation of semiosis from referentiality.

Today, as the new vortices of power are shaped by the instant electric interdependence of all men on this planet, the visual factor in social organization and in personal experience recedes, and money begins to be less and less a means of storing or exchanging work and skill. Automation, which is electronic, does not represent physical work so much as programmed knowledge. As work is replaced by the sheer movement of information, money as a store of work merges with the informational forms of credit and credit card.[14]

The loss of physicality of money is part of the general process of abstraction, the all-encompassing tendency of capitalism.

Retracing the history of money from exchange commodity, to representative money, to standard value, to electronic abstraction, McLuhan writes:

> [...] the Gutenberg technology created a vast new republic of letters, and stirred great confusion about the boundaries between the realms of literature and life. Representative money, based on print technology, created new speedy dimensions of credit that were quite inconsistent with the inert mass of bullion and of commodity money. Yet all efforts were bent to make the speedy new money behave like the slow bullion coach. J. M. Keynes stated this policy in *A Treatise on Money*:
>
> > Thus the long age of Commodity Money has at last passed finally away before the age of Representative Money. Gold has ceased to be a coin, a hoard, a tangible claim to wealth, of which the value cannot slip away so long as the hand of the individual clutches the material stuff. It has become a much more abstract thing—just a standard of value; and it only keeps this nominal status by being handed round from time to time in quite small quantities amongst a group of Central Banks.[15]

Marx's theory of value was based on the concept of abstract work. Because it was the source and the measure of value, work had to sever its relation to the concrete usefulness of its activity and product. Concrete usefulness did not matter from the point of view of valorization. The process of abstraction at the core of the capitalist capture, or subsumption, of work, implied abstraction from the need for the concreteness of products: the referent was erased. Baudrillard speaks of the relation between signification and language in the same vein:

> The rational, referential, historical and functional machines of consciousness correspond to industrial machines. The aleatory, non-referential, transferential, indeterminate and floating machines of the unconscious respond to the aleatory machines of the code. [...] The systemic strategy is merely to invoke a number of floating values in this hyperreality. This is true of the unconscious as it is of money and theories. Value rules according to the indiscernible order of generation by means of models, according to the infinite chains of simulation.[16]

The crucial point of Baudrillard's critique is the end of referentiality and the (in)determination of value. In the sphere of the market, things are not considered from the point of view of their concrete usefulness, but from that of their exchangeability and exchange value. Similarly, in the sphere of communication, language is traded and valued as performance. Effectiveness, not truth-value, is the rule for language in the sphere of communication. Pragmatics, not hermeneutics, is the methodology for understanding social communication, particularly in the age of new media.

Retracing the process of the loss of reference in both semiotics and economics, Baudrillard speaks of the emancipation of the sign.

A revolution has put an end to this "classical" economics of value, a revolution of value itself, which carries value beyond its commodity form into its radical form.

This revolution consists in the dislocation of the two aspects of the law of value, which were thought to be coherent and eternally bound as if by a natural law. *Referential value is annihilated, giving the structural play of value the upper hand.* The structural dimension becomes autonomous by excluding the referential dimension, and is instituted upon the death of reference. [...] from now on signs are exchanged against each other rather than against the real (it is not that they just happen to be exchanged for each other, they do so *on condition* that they are no longer exchanged against the real). The emancipation of the sign.[17]

The emancipation of the sign from the referential function may be seen as the general trend of late modernity, the prevailing tendency in literature and art, as well as in science and politics.

In the passage from romantic realism to symbolist trans-realism, a new space for poetic praxis was opened, and the emancipation of the word from its referential task was the main gateway to the new semiotic laboratory that was art in the century of the avant-garde.

The emancipation of money—the financial sign—from the industrial production of things follows the same semiotic procedure, from referential to self-referential signification.

Dark Pools: Financial Abstraction

In Marx's writings, abstraction was viewed as the main trend of capitalism, the general effect of capitalism on human activity. Nevertheless, in the process of industrial production, something useful had to be produced if the capitalist wanted to sell his merchandise and thus increase the capital that he invested at the beginning of the production process. The use value of the worker's product was only a step toward the real thing, which was surplus value. So the capitalist did not care whether his work produced chickens, or books, or cars ... He cared only about this: how much value could his work produce in a given unit of time. Nevertheless, something useful had to be produced in the process.

This is no longer true in the present conditions of financial capitalism, since the increase of monetary capital does not need to go through the production of useful goods. The capitalist is no longer obliged to invest money in a useful good in order to acquire more money at the end of the exchange. Financial virtualization has made a new cycle of valorization possible: money can become more money, skipping the passage of the production of useful goods.

Financial virtualization is the ultimate step in the transition to the form of semiocapital. In this sphere, two new levels of abstraction appear, as developments of the abstraction of labor that Marx wrote about.

Digital abstraction adds a second layer to capitalist abstraction. Transformation and production no longer occur in the field of bodies, of material manipulation, but in the field of pure self-referential interaction between informational machines. Information takes the place of things, and the body is cancelled from the field of communication.

Then there is a third level of abstraction, which is financial abstraction. Finance—which once upon a time used to be the sphere where productive projects could meet capital, and where capital could meet productive projects—has been untied from the need for production. The process of the valorization of capital, that is, that of increasing the money invested, no longer passes through the step of producing use value, or even that of producing physical or semiotic goods.

In the old industrial economy described by Marx, the goal of production was already the valorization of capital, through the extraction of surplus value from labor. But, in order to produce value, the capitalist was still required to exchange useful things, so he was required to produce cars, and books, and bread.

When the referent is cancelled, when profit is made possible through the mere circulation of money, the production of cars, books, and bread becomes superfluous. The accumulation of abstract value is made possible through the subjection of human beings to debt, and through the predation of existing resources. The destruction of the real world starts from this separation of valorization from the production of useful things, and from the self-replication of value in the financial field. The separation of value from the referent leads to the destruction of the existing world. This is exactly what is happening under the cover of the so-called financial crisis, which is not a crisis at all, but the transition to self-referential financial capitalism.

In the book *Data Trash*, Arthur Kroker and Michael Weinstein write that in the field of digital acceleration, more information means less meaning, because meaning slows down the circulation of information. In the sphere of the digital economy, the faster information circulates, the faster value is accumulated. But meaning slows down this process, since meaning needs time to be produced and to be elaborated and understood. So the acceleration of information flow implies the elimination of meaning.[18]

In the sphere of the financial economy, the acceleration of financial circulation and valorization imply the elimination of the concrete usefulness of products, no matter whether they are material or immaterial, industrial or semiotic. The process of realization of capital, namely the exchange of goods for money, was obviously slowing the pace of monetary accumulation. Virtual technology has created the possibility of skipping this slow passage through concrete, meaningful and useful goods. This is what Christian Marazzi calls the "becoming rent of profit."

The increase of the abstract entity that is money is expedited by virtualization, and the virtual circulation of abstract value, which is not even money but algorithms and electronic impulses, has been accelerated and sped up to the point of totally escaping beyond the possibility of human understanding and—obviously—of political control.

In the book *Dark Pool*, the journalist Scott Patterson speaks of the extreme acceleration of financial technology. The book is an account of how global markets have been hijacked by trading robots.

Patterson speaks of *algo wars* to refer to the way fiber-optic cables link financial markets at ever-increasing speeds, and to the fact that these cables are now being superseded by even faster microwave stations relaying high-speed financial trade. In the past, economic exchange used to occur in a physical place, where people would come together to buy or to sell, hoping to garner the best price.

In the last twenty years, computers, electronic exchanges, dark pools, flash orders, multiple exchanges, alternative trading venues, direct access brokers, OTC derivatives, and high-frequency traders have totally changed the financial landscape, and in particular, the relation between human operators and self-directing algorithmic automatons. Patterson predicts that the high-frequency trend will continue. The more you remove the reference to physical things, to physical resources, and to the body, the more you can accelerate the circulation of financial flows.

In Greek, *parthenos* means *virgin*. Jesus Christ was created by parthenogenesis. The Virgin Mary gave birth to her son without any relationship to the reality of sex. Similarly, financial economy, like conceptual art, is a parthenogenetic process. The monetization and financialization of the economy represent a parthenogenetic transformation of the creation of value. Value no longer emerges from a physical relationship between work and things, but rather from the infinite self-replication of virtual exchanges of nothing with nothing, whose outcome is more money.

Digital abstraction virtualizes the physical act of meeting, and the manipulation of things. These new levels of abstraction concern not only the labor process, but they tend to encompass all spaces of social life. Digitalization and financialization are transforming the very fabric of the social body, and inducing mutations within it.

In the Realm of Floating Values

Chomsky's structural theory is based on the idea that linguistic signs can be exchanged thanks to a bank of shared structures, that is, a common cognitive competency that makes the exchange possible. In the same way as money is a general equivalent, a universal translator of different goods, so is language. We can exchange everything with money; we can exchange everything with words.

But money is also a tool to force people to do something, and in the sphere of financial capitalism, it is increasingly a pragmatic act of self-expansion. The present configuration of the economic landscape emphasizes this side of money, stressing its characteristic as a pragmatic tool to compel people to accept any kind of work, to be submissive, to suffer exploitation, humiliation, and violence.

In the sphere of financial capitalism, money is less an indicator than a factor of mobilization. Look at the reality of debt, look at the awful effects of submission, impoverishment, and exploitation that debt is provoking in the body of society. Debt is a weapon against social autonomy, a transformation of money into blackmail. Since neoliberal fanatics have destroyed public education, and the costs of private school are skyrocketing, young people are forced to borrow money in order to pay for their education. As soon as they graduate from the university, they have to start paying back their debt, and they are forced to accept any precarious job, and to suffer all forms of blackmail.

Money, which was supposed to be the measure of value, has been turned into a tool for psychic and social subjugation. A metaphysical debt links money, language, and guilt. Debt is guilt, and as guilt, it is entering the domain of the unconscious, shaping language according to structures of power and submission.

The pragmatic effect of financial language crystallizes into a linguistic structure that everybody is forced to share. For us to understand each other, we have to go to the bank of universal translation, since understanding is based on a common standard, a conventional measure that is deposited in said bank. And the translator is both the fabricator of meaning and the owner of the relation between meaning and life.

Guattari breaks apart this idea of a universal exchange of structural signifiers, and opens a vista onto a different landscape. For him, language is essentially the pragmatics of communication, the creation of meanings that did not exist before the act of communication. I want to stress here the analogy between this kind of pragmatics of communication and the financial creation of value out of nothing, out of pure exchange and virtual activation. In *The*

Machinic Unconscious, Guattari speaks of the ritornello in order to define the relation between a singularity and the cosmos. Language is the creation of singular ritornellos that concatenate with the cosmos, as long as they are crystallized by structures of power and of cognitive automation.

The Machinic Unconscious was an attempt to get rid of the Chomskian idea that language is governed by the structure of the universal mind. There is no universal mind, no universal structure of language, but there are a-signifying signs that produce meaning thanks to the concatenation with other a-signifying signs. Meaning does not inhere in a universal mind-grammar but in the sliding relation between a voice, a listener, and a context. Meaning is intention, agreement, conflict, and desire.

When we enter the flow of communication, we are not interpreting signs in relation to a structure, we are producing, and there is no structure before the act of concatenation. Language is the product of linguistic activity, and this activity is a constant variation of existing expectations of meaning.

What happened after Arthur Rimbaud declared the disordering, or deregulation, of all the senses (*le dérèglement de tous les sens*).[19]

The impressionists declared they did not want to show the thing, but wanted to show its impression. The symbolists invited readers to forget about the referent. Symbolist poetry wanted to be evocation. Breaking the fixed relation between the referent, interpretation, and structure, symbolist poetry re-invented the relation between words and things. No more representation, but evocation, deterritorialization of meaning, epiphany, and simulation.

This process of the de-referentialization of language—emancipation of the linguistic sign from its referent—that has been the mark of poetic and artistic experimentation with language over the course of the twentieth century, shows an interesting similarity with the transformation in the relation between the economy and monetary exchange.

On August 15, 1971, President Nixon announced dramatic changes in economic policy. More particularly, he ended the Bretton Woods international monetary system. The Bretton Woods system, created at the end of World War II, involved fixed exchange rates with the US dollar as the key currency—but also a role for gold linked to the dollar at $35 an ounce. The system began to falter in the 1960s because of an excess of dollars flowing out of the US, and which foreign central banks had to absorb. All of this was ended unilaterally by Nixon's decision. After a brief attempt to create a modified fixed exchange rate system, the world moved to flexible rates.

Nixon's decision can be viewed as an act of *de-referentialization* in the realm of monetary economy. Breaking the Bretton Woods agreements, the American president declared that the dollar had no referent, and that its value was decided by an act of language. This was the beginning of the long-lasting process of financialization of the economy, based on the emancipation of the financial dynamic from any conventional standard and from any economic reality. The neoliberal offensive started at that very moment, when it decided upon the arbitrary assertion of the value of the dollar outside of any conventional standard. For the neoliberal school of the Chicago Boys money created reality, just as words had created reality for the symbolist poets.

The hypertrophic self-replication of debt began then. Financial economy no longer dealt in producing things, but rather evoked the world through the circulation of money.

Some years after the deregulation of the international monetary system, Jean Baudrillard wrote *Symbolic Exchange and Death*, where he announced that the economy had abandoned the old law of the determination of value, and that the referent for linguistic and economic exchange had been dissolved.

> The reality principle corresponded to a certain stage of the law of value. Today the whole system is swamped by indeterminacy, and every reality is absorbed by the hyperreality of the code and simulation. The principle of simulation governs us now, rather than the outdated reality principle. We *feed* on those forms whose finalities have disappeared. [...]
>
> The systemic strategy is merely to invoke a number of floating values in this hyperreality. This is as true of the unconscious as it is of money and theories. Value rules according to the indiscernible order of generation by means of models, according to the infinite chains of simulation.[20]

Futurist Mythology

Futurists rejected symbolism as a faint, languid, and womanly style. Nevertheless, the futurist mythology of the omnipotence of the constructive act was based on the symbolist removal of reality as a referent from the sphere of poetical creation. "Forget about the referent," says the symbolist, to which the futurist adds, "Nothing pre-exists the semiotic activity of the inventor, the destroyer-constructor, the artist. Only the future exists, created by the annihilation of past."

On February 20th, 1909, Filippo Tommaso Marinetti published the first futurist manifesto. In the same year, in an automobile factory in Detroit, Henry Ford put into operation the first assembly line, the technological system that best defined the age of industrial massification. Both events can be considered to inaugurate a century in which society invested psychic and cultural energy in the future dimension of ever-expanding wealth and knowledge. Speed and the cult of the machine were the values emphasized by the *Futurist Manifesto*.

It is significant that the futurist movement surfaced both in Italy—and in Russia. These two countries shared common conditions of social and economic backwardness, a scant development of industrial production, a marginal expansion of the bourgeois class, a heavy reliance on the cultural and religious models of the past, and an attraction for foreign culture, especially French, in urban intellectuals.

Underlying the whole edifice of modernity was the mythology of speed. Progress was in fact based on the intensification of the productivity of labor.

The Futurist Manifesto asserted the aesthetic value of the machine. The machine par excellence was the speed machine— the car, the airplane—tools that made it possible to mobilize the social body.

Futurism exalted the machine as an external object, visible in the city landscape, but in the twenty-first century, the machine has now become internalized. We are no longer focused on the external machine; the contemporary *info-machine* now intersects with the social nervous system, the *bio-machine* interacts with the genetic becoming of the human organism. Digital and biotechnologies have turned the external machine of iron and steel into an internalized and recombining machine. The bio-info machine is no longer separable from the body and the mind, because it is no longer an external tool, but instead has become an internal transformer of the body and of the mind, a linguistic and cognitive enhancer. The nano-machine will cause the human brain and the linguistic ability to produce and communicate to mutate.

We are the machine.

In the mechanical era, the machine stood in front of the body, and transformed human behavior, enhancing bodily potency by means of external imposition. The assembly line, for instance, although it improved and increased the productive power of the laborer, did not modify his or her physical organism and cognition. Now the machine is no longer in front of the body but inside the body-mind. Because of this transformation, the nature of political power has changed. When the machine was external, the state had to regulate the body that enforced the law. Agencies of repression were mobilized in order to force the conscious organisms to submit to that rhythm without rebelling. Now, political domination has become internalized and undistinguishable from the machine itself.

In the connective sphere, the machine is a difference of information. It is no longer an exterior device but a system of cognitive automatisms and internal necessity.

A hundred years after the publication of the *Futurist Manifesto*, speed itself has been internalized. The colonization of the mind and of perception is based on an interior acceleration in the perception of time.

Language, Rhythm, Respiration

A long time ago, I happened to take part in an action with the Living Theater. In an old Italian theater, some hundred people met for a collective mantra: the collective emission of sound, shared breathing, and a vocal wave flowing from one mouth to the next, from one body to the next.

Yogic wisdom conceives of individual breathing (*atman*) as the relation of the organism with cosmic breath (*prana*).

Physical organisms interact with the natural environment, the city, the workplace, and with a polluted atmosphere. Psychic organisms also interact with the environment, namely with the infosphere where info-stimuli circulate.

Semiotic flows spread by the media through the infosphere are polluting the psychosphere and provoking disharmony in psychic breathing. Fear, anxiety, panic, and depression are symptoms of the illness provoked by this kind of pollution.

Conjunction can be compared to breathing together, as it implies the exchange and transmission of material substance, that is, the physical matter contained in the air we breathe, or the semiotic matter conveyed by signs. The search for a common rhythm, the tentative interpretation of bodily and semiotic nuances, and the non-verbal disambiguation of verbal signs are part of conjunctive communication.

Sensibility is the possibility of entering into a relation with entities that do not speak our language and that are composed of substances that differ from ours.

Sensibility is the ability to harmonize with the heterogeneous rhizome.

Collective assemblages of enunciation function directly within *machinic assemblages*; it is not impossible to make a radical break between regimes of signs and their objects. [...] The orchid deterritorializes by forming an image, a tracing of a wasp; but the wasp reterritorializes on that image. The wasp is nevertheless deterritorialized, becoming a piece in the orchid's reproductive apparatus. But it reterritorializes the orchid by transporting its pollen. Wasp and orchid, as heterogeneous elements, form a rhizome.[21]

On the ontological, teleological, or even the physical plane, the wasp and the orchid are not homogeneous. They even belong to different natural realms. But this does not prevent them from working together in the sense of becoming a concatenation, and in so doing generating something that did not exist before. "Be, be, be!" is the metaphysical scream that dominates hierarchical thought. Rhizomatic thought replies: "Concatenate, concatenate, concatenate!"

The principle of becoming lies in conjunctive concatenation:

> [...] a becoming-wasp of the orchid and a becoming-orchid of the wasp. Each of these becomings brings about the deterritorialization of one term and the reterritorialization of the other; the two becomings interlink and form relays in a circulation of intensities pushing the deterritorialization ever further. There is neither imitation nor resemblance, only an exploding of two heterogeneous series on the line of flight composed by a common rhizome that can no longer be attributed to or subjugated by anything signifying. Rémy Chauvin expresses it well: "the *aparallel evolution* of two beings that have absolutely nothing to do with each other."[22]

The mutation that is currently investing the social and linguistic organism can be essentially described as a transition from the sphere of conjunction to the sphere of connection. In the shift, a change occurs. While conjunctive communication is a tentative approach to the intentions of meaning of a body that sends ambiguous messages whose interpretation is an object of negotiation and uncertainty, connective communication implies and presupposes a perfectly unambiguous interaction between agents of signification that are syntactically compatible. If you want to figure out what conjunctive communication is, just imagine two people engaged in courting, an activity that involved desire, shyness, ambiguity, innuendoes of intuition, and infinite layers of (mis)understanding.

If you want to understand what connective communication is, think of the syntactic overlapping and semantic identification between two strings of information. Connection is the interaction between syntactic machines that have the same format. When human beings want to take part in a connection, they must previously accept the syntactic reduction of the contents of their exchange to the format of the machines that are carrying their signs.

Composition and Recombination

Composition involves the mixture of different chemical substances. Respiration implies a process of composition, of blending, and of changing the organism. If you happen to breathe poisonous substances, your organism will get sick. Social composition is the process of cultural contamination between conscious and sensible organisms that are more or less different, as they share the same ability to understand what is not exactly inter-exchangeable, but what can be exchanged through small (or large) transformations in the organisms themselves.

In the parlance of Italian *post-operaismo* (post-workerism), the notion of social (class) re-composition plays an important role, as it refers to the process that underlies political solidarity, and the creation of cultural and psychological conditions for the social autonomy of workers from capitalist rule. In the process of re-composition, the entire range of social life is involved, implicating expectations, coalescing lifestyles, national or ethnic conflicts, psychological empathy, mythologies, cultural traditions, and so on.

Solidarity is a pre-condition for political organization and social struggle. The history of class conflicts, of victories and defeats cannot be explained without referring to the degree of solidarity that workers have been able to establish. Solidarity is not based on ethical or ideological issues. It depends on the features of the relations between individuals in time and in space. The material foundation of solidarity is the perception of the continuity of bodies, and the immediate understanding of the consistency of my interest with yours.

Solidarity has to do with conspiracy, which means, precisely, breathing together (from the Latin, *conspirare*, from *con*, *together with*, and *spirare*, *to breathe*).

During the first part of the twentieth century, when industrial workers were fighting worldwide against imperialist oppression and capitalist exploitation, the communist conspiracy was the psychic and cultural energy that made solidarity possible within the social body of the industrial working class, notwithstanding the authoritarian reality of communist parties and the unspeakable violence of communist states.

The process of re-composition is endangered and jeopardized by the technical and social restructuring of social machinery. In the last decades of the past century, the neoliberal offensive, linked with the technical deterritorialization provoked by globalization, destroyed previous forms of the social organization of workers, and started a process of de-composition leading to the precariousness of labor.

Precarity has provoked a loss of solidarity and the disaggregation of the social composition of work. The virtualization of production, enabling delocalization, is a complementary cause of loss of solidarity. Precarious labor triggers social competition between workers. As work is transformed into information, virtualization jeopardizes affective relations between people involved in the labor process.

The product of deterritorialized work is recombined by the network.

Recombination, which is totally different from recomposition, implies compatibility and functional inter-operationality between deterritorialized working bodies. In order to circulate in the network, language has to be made compatible with the code.

The human beings involved in the productive process, precarious cooperators whose social composition is fragmentary, are transformed into fractals, perfectly recombinable segments of a modular flow of information.

The social body is fragmented, and breath is broken and subjected to the rhythms of the virtual machine. Furthermore, the fractal fragmentation of labor is parallel and complementary to the fractalization of financial capital, the continuous recombination of abstract financial substance, leading to virtual fragments of disembodied depersonalized capital.

5

Avatars of the General Intellect

The Matrix and the Cloud

In my view, Lana and Andy Wachowski's films—*The Matrix* and *Cloud Atlas*—are visual meditations about determinism and freedom, about neuroplasticity and the fabric of time. Can the Matrix capture cognition and sensibility, when we know that cognition and sensibility are as impossible to map as it is impossible to map a cloud?

The all-encompassing web of algorithmic automatons pervading the sphere of financial capitalism is an attempt at mapping and subjecting the general intellect. This attempt may succeed inasmuch as the general intellect can be reduced to a system of operational functions, of logical implications, and of technological interactions. However, it may not succeed since the general intellect has a body, and the body of the general intellect is the body of uncountable cognitive workers who live in conditions of precarious pay, in stressful conditions of competition, exploitation, and nervous hyper-stimulation.

Here lies the weak point of the Matrix, here is the only way out of the process of the final domination of the social brain, the neurototalitarianism that is perceivable today as a deadly, and impending possibility.

The cognitarian body—the cognitive work force subsumed by the linguistic machine—is composed of the individual existence of millions of people who are sitting in front of the networked screen, the virtual assembly line of semiocapitalism. Consequently, this body is not entirely reducible to the Matrix, because it is not only intellect, but sensibility as well. The general intellect has a body, and this body is a sensible-sensitive body that feels pleasure and pain, as long as it is not subjected to utter anesthesia. The body of the general intellect is a cloud, an ever-variable vibration of emotions, expectations, fears, desires, and exhaustion. The cloud cannot be mapped, as David Mitchell suggests in his novel, and Lana and Andy Wachowski in their film.

The cloud is the irreducibility of the psychosphere to global determinism and ultimate traceability. Sensibility is the excess, the surplus of vibrational life that cannot be translated into algorithm.

Abstract Labor and General Intellect in Marx

Let's start from the concept of abstract labor. With this expression, Marx refers to value as the crystallization of labor time, and refers to labor as time materialized in value. What capital must mobilize is not the concrete ability to produce useful things, but the abstract ability of time without quality to generate value.

> The indifference as to the particular kind of labor implies the existence of highly developed aggregates of different species of concrete labor, none of which is any longer the predominant one. So the most general abstractions commonly arise only where there is the highest concrete development [...]. On the other hand, this abstraction of labor is only the result of a concrete aggregate of different

kinds of labor. The indifference to the particular kind of labor corresponds to a form of society in which individuals pass with ease from one kind of work to another, which makes it immaterial to them what particular kind of work may fall to their share.[1]

The abstraction of labor progressively expands to all possible forms of social activity. The final point of this process is the subsumption of mental activity itself into the sphere of capital valorization, and abstraction of mental activity itself.

> Capital necessarily tends towards an increase in the productivity of labor and as great a diminution as possible in necessary labor. This tendency is realized by means of the transformation of the instrument of labor into the machine [...]. The value objectified in machinery appears as a prerequisite, opposed to which the valorizing power of the individual worker disappears, since it has become infinitely small.[2]

Thanks to the accumulation of science and the general forces of the social intellect, physical labor becomes superfluous. The tendency of capital is to eliminate human labor as much as possible, in order to replace it through the technological use of science. But capital simultaneously needs to exploit human labor, as abstract value is only generated by it.

> Nature builds no machines, no locomotives, railways, electric telegraphs, self-acting mules, etc. These are products of human industry; natural material transformed into the organs of the human will over nature, or of human participation in nature. They are *organs of the human brain, created by the human hand*; the

power of knowledge, objectified. The development of fixed capital indicates to what degree social knowledge has become a *direct force of production*, and to what degree, hence, the conditions of the process of social life itself have come under the control of the general intellect and been transformed in accordance with it. To what degree powers of social production have been produced, not only in the form of knowledge, but also as immediate organs of social practice, of the real life process.[3]

The conceptual development of this tendency takes the productive system virtually outside the paradigmatic orbit of capitalism. By replacing work with machines that in turn have the ability to produce more machines replacing human work, capital reduces the time of labor that is needed for social reproduction. The social need of labor time tends to zero. This is why Marx says that capitalism is actively working to its own dissolution. But in order to counteract its own dissolution, capital is also working against this tendency, producing scarcity and need, and destroying the products of work in many ways, through war, crises of overproduction, and financial collapse.

Thanks to technology, productivity has significantly increased. In one hour, we can produce the same amount of goods that used to require the work of an entire day. However, the time captured and subjected to work—after a temporary decrease in the second part of the twentieth century—is now increasing again, absorbing most of the time of social life.

One may say that social civilization and human progress have essentially been about the emancipation of the time of life from the obligation to work. When working time decreases, people are allowed to dedicate their energies to mutual attention, self-care, education, and pleasure. The production of useful things and useful services does not decline, but rather increases when we are free from the obligation of abstract work.

After a few decades of decrease of working time, which happened to coincide with the spread of progressive culture and social movements for self-determination, neoliberal ideology launched a global campaign for the long-lasting reduction of real wages in order to force people to work more.

The increase of working time and the intensification of productivity were not aimed at improving people's lives, but rather at implementing economic growth, which means the accumulation of capital. Subjecting social energies to the domination of money is the neoliberal way to reassess the primacy of accumulation over social wellbeing.

What Marx underestimated in the visionary prediction contained in the *Fragment on Machines* is the cultural force of the paradigm based on accumulation, the metaphysical greed that transforms the life of the world into a mere tool for economic expansion.

Capital semiotizes the potentialities of the general intellect according to a monetary paradigm that entangles and perverts the ability of intellectual labor to increase the production of useful things while simultaneously reducing working time.

Here is a paradox that Marx had been able to sense, but not to fully clarify.

> Capital has quite unintentionally reduced human labor, the expenditure of energy, to a minimum. This will be to the advantage of emancipated labor, and is the condition of its emancipation. [...][4]
>
> As soon as labor, in its direct form, has ceased to be the direct form of wealth, then labor time ceases, and must cease, to be its

standard of measurement, and thus exchange value must cease to be the measurement of use value. The surplus labor of the masses has ceased to be a condition for the development of wealth in general; in the same way that the non-labor of the few has ceased to be a condition for the development of the general powers of the human mind. Production based on exchange value therefore falls apart, and the immediate process of material production finds itself stripped of its impoverished, antagonistic form. Individuals are then in a position to develop freely. It is no longer a question of reducing the necessary labor time in order to create surplus labor, but of reducing the necessary labor of society to a minimum. The counterpart of this reduction is that all members of society can develop their education in the arts, sciences, etc., thanks to the free time and means available to all.[5]

The economic system of capital, acting as a general semiotic cage, forbids the possibility of emancipation from labor, while it simultaneously expands the capability of the general intellect to replace human work with technology.

Capital is itself contradiction in action, since it makes an effort to reduce labor time to the minimum, while at the same time establishing labor time as the sole measurement and source of wealth. Thus it diminishes labor time in its *necessary* form, in order to increase its *superfluous* form; therefore it increasingly establishes superfluous labor time as condition [...] for necessary labor time. On the one hand it calls into life all the forces of science and nature [...] in order to create wealth which is relatively independent of the labor time utilized. On the other hand it attempts to measure, in terms of labor time, the vast social forces thus created and imprisons them within the narrow limits that are required in order to retain the value already created *as* value.[6]

These pages are a conceptual map prefiguring the social and technological development of twentieth-century history, from the acceleration of mechanical machinery to the digitalization of global production in the last decades of the century.

When Marx spoke of capital as a contradiction in process, he prefigured the astonishing history of the twentieth century, when capital itself, led by an instinct to conserve its own social and economic model, destroyed the very potentialities created within the technical domain. When he spoke of the development of creative, artistic, and scientific faculties, Marx foretold the intellectualization of labor that is clearly visible today. After intensifying technological progress, capital turns into a semiotic tangle.

Figures of the Modern Intellectual

The word *intellectual* has lost much of its meaning today, but during the course of the twentieth century this word was crucial in the fields of ethics and politics. In late modernity, the nature of intellectual labor dramatically changed, as it was progressively absorbed within economic production. When digital technologies made it possible to connect individual fragments of cognition with semiotic production, intellectual labor was captured and subsumed by the cycle of the production of value.

The Enlightenment did not define the intellectual by his or her social condition, but as the incarnation of ideology, as the source of the universal system of values. In the sphere of the Enlightenment, the intellectual was the founder and the guarantee of the realization of universal principles, the respect for the rights of man, and the universality of law.

In the context of Kantian thought, the intellectual emerged as a transcendent figure, whose activity was independent of social experience, or in any case, not socially determined in its cognitive or ethical decisions. The intellectual appeared then as the bearer of a universal rationality, abstractly human, and in that sense the intellectual could be considered the historical determination of the Kantian "I think" (*Ich denke*).

In that sense, the intellectual was the guarantor of democracy. Democracy cannot descend from cultural origins, from some kind of belonging, but only from the unlimited horizon of choice, from the possibility of access and of citizenship for every person as a semiotic agent, as a subject who exchanges signs in order to access universal rationality. The figure of the intellectual in this sense sets itself up against the romantic figure of the *volk*, or rather withdraws itself from it. Universal thought, from which the modern adventure of democracy was born, evades the territoriality of culture. Democracy cannot carry the imprint of a culture, of a people, or of a tradition; it must be a game with foundations, inventions, or conventions, and without the affirmation of belonging.

Significantly different is the point of view of the revolutionary intellectual that affirms itself with historical-dialectical thought. In the eleventh thesis on Feuerbach, Marx writes, referring to the role that knowledge must play in the historical process: "Philosophers have hitherto only interpreted the world in various ways; the point is to change it."

The Marxist intellectual conceives of himself or herself as an instrument of the historical process of achieving a classless society. According to Marx, thought is historically effective only when it recognizes in the working class the horizon of action. The communist project views theory as a material power, and knowledge as an instrument for changing the world. Only inasmuch as he participates in the struggle for the abolition of the exploitation of labor does the intellectual become the bearer of a universal mission. The intellectual has nothing to do with the people (*volk*), in this vision, because the people is the territorialized figure of belonging, the predominance of *kultur* with respect to reason, the preeminence of the root with respect to finality. On the contrary, the working class does not belong to any territory, to any culture, and its mental horizon is that of a universally exploited class, striving towards a universal task of freeing from exploitation.

The role of intellectuals is central in the framework of communist revolutionary thought, particularly in Lenin's conception. In *What Is to Be Done?*, Lenin argues that intellectuals are not a social class, they have no specific social interests to uphold. They generally come from the upper classes or the petite bourgeoisie, and carry out *purely intellectual* choices, turning themselves into intermediaries and organizers of a revolutionary consciousness descending from philosophical thought. In this sense intellectuals are similar to the pure becoming of the spirit, to the Hegelian unfolding of self-consciousness. On the other hand, workers, still bearers of social interests, can shift from a purely economic existence (the Hegelian *an sich* of the immediacy of social being) to politically conscious activity (the *für sich* of self-consciousness), only through the political form of the party, which incarnates and transmits the philosophical legacy to the masses. In Leninist parlance, it is possible to speak of the proletariat as heir to German classical philosophy: thanks to the workers' struggle, the historical realization of the dialectical horizon, the final point of German philosophical development becomes possible.

In Gramsci, the reflection on intellectuals has the connotation of a social analysis, and approaches a materialist formulation of the organic character of the relationship between intellectuals and the working class. The collective dimension of intellectual activity is nonetheless identified in the party, which is defined as the collective intellectual. The intellectual of the modernist tradition can access the collective and political dimension only by adhering to the party. In the second part of the twentieth century, mass education and techno-scientific change reframed the role of intellectuals. They are no longer independent from production, no longer free individualities that take upon themselves the task of a purely ethical and freely cognitive choice, but a massive social subjectivity that tends to become an integral part of the general productive process. Paolo Virno uses the term *mass intellectuality* in this context. In his view, the emergence of the student movement in the sixties was the conscious expression of this new figure of mass intellectuality.

During the century of communist revolutions, the Leninist tradition disregarded the notion of the general intellect, yet in the light of post-industrial transformation it emerges as a central force. As digital technologies and the creation of the global network redefine the overall social process around the general intellect at work, the Leninist conception of the party and even the Gramscian notion of the organic intellectual lose consistency.

Cybertime and the Expansion of Capitalism

Rosa Luxemburg argued that capitalism is intimately pushed towards a process of continual expansion. Imperialism is the political, economic, and military expression of this need for continual expansion that brings capital to continually extend its domain.

But what happens when every inch of planetary territory has been subjected to the rule of the capitalist economy, and every object of daily life has been transformed into a commodity? In late modernity, capitalism seems to have exhausted every possibility for further expansion. For a certain period, the conquest of extraterrestrial space seemed to be a new direction of development for capitalist expansion. Subsequently we saw that the direction of development was above all the conquest of internal space, the interior world, the space of the mind, of the soul, and the space of time.

The colonization of time has been a fundamental objective of the development of capitalism during the modern era. The anthropological mutation that capitalism produced in the human mind and in daily life has been above all a transformation in the perception of time.

Yet with the spread of digital technologies, which allow absolute acceleration, something new occurs. Time becomes the primary battlefield, as it is the space of the mind, mind-time, cybertime.

I have here introduced a distinction between the concept of cyberspace and the concept of cybertime. Cyberspace is the sphere of connection of innumerable human and machinic sources of enunciation, the sphere of connection between minds and machines in unlimited expansion. This sphere can grow indefinitely, because it is the point of intersection between the organic body and the inorganic body of the electronic machine.

But cybertime is the organic side of the process, and its expansion is limited by organic factors. The human brain's capacity to elaborate can be expanded with drugs, with training and attention, but the organic brain has limits that are connected to the emotional, and sensitive dimension of the conscious organism.

Cybertime is not a purely extensible dimension, because it is connected with the intensity of experience. The objective sphere of cyberspace expands at the speed of digital replication, but the subjective nucleus of cybertime evolves at a slower rhythm, the rhythm of corporeality, of pleasure, and of suffering.

The technical composition of the world changes, but cognitive appropriation and psychic reactivity do not follow in a linear manner. The mutation of the technological environment is much more rapid than the changes in cultural habits and cognitive models.

The stratum of the infosphere grows progressively thicker and denser, and informational stimulus invades every atom of human attention.

Cyberspace grows without limits, yet mental time is not infinite. The subjective nucleus of cybertime follows the slow rhythm of organic matter. We can augment the time of exposure of the organism to information, but experience cannot be intensified beyond certain limits.

Beyond certain limits, the acceleration of experience provokes a reduced consciousness of stimulus, a loss of intensity that concerns the aesthetic sphere, that of sensibility, and also the sphere of ethics. The experience of the other becomes awkward, even painful, as the other becomes part of an uninterrupted and frenetic stimulus, and loses its singularity, intensity, and beauty. The consequence is a reduction of curiosity, and an increase in stress, aggressiveness, anxiety, and fear.

The acceleration of the infosphere produces an impoverishment of experience, because we are exposed to a growing mass of stimuli that we cannot elaborate intensively, or deeply know and perceive.

More information, less meaning. More stimuli, less pleasure. Sensibility occurs within time. Sensuality develops in slowness, and the space of information is too extensive and fast for sensuality to elucidate intensively and deeply. The fundamental crux of the contemporary mutation lies at the point of intersection between electronic cyberspace and organic cybertime.

The social brain is subjected to the invasion of video-electronic flows, and experiences the superimposition of digital code over the codes of recognition and identification that mold organic cultures.

The acceleration produced by network technologies, and the precarious conditions of cognitive labor provoke a pathogenic effect of saturation of the time for attention. The pathology of cognitive labor is the new condition of alienation, a prerequisite to the cognitarian rebellion and possibly to the recomposition of the body of the general intellect.

The Intellectual, the Merchant, and the Warrior

> There's a time when the operation of the machine becomes so odious, makes you so sick at heart, that you can't take part, you can't even passively take part. And you've got to put your bodies upon the gears and upon the wheels, upon the levers, upon all the apparatus, and you've got to make it stop. And you've got to indicate to the people who run it, to the people who own it, that unless you're free, the machine will be prevented from working at all.[7]

These words, pronounced by Mario Savio at Berkeley's Sproul Hall in the year 1964, can be considered the beginning of the movement that shook the world in the 1960s and that peaked in 1968. The student movement was originally motivated by the understanding that knowledge submits to the military system, particularly to the criminal war that the United States was waging in Indochina.

For the students of the Free Speech Movement, the university was an instrument of the war politics of the American government and of the overall capitalist machine.

The place where the student's revolt began is the same as where—years later—the new industry of electronics and computing thrived. In the book *From Counterculture to Cyberculture*, Fred Turner writes:

> Thirty years later, the same aspects of computing that threatened to dehumanize the students of the Free Speech Movement promised to liberate the users of the Internet. On February 8th, 1996, John Perry Barlow, an information technology journalist and pundit, and a former lyricist for the house band of the San Francisco LSD scene, the Grateful Dead, found himself at his laptop computer in Davos, Switzerland. While attending the World Economic Forum, an international summit of politicians and corporate executives, he had watched the American Congress pass the *Telecommunications Act*, and with it a rider called the *Communicative Decency Act*, which aimed to restrict pornography on the Internet. Incensed by what he perceived to be the rider's threat to free speech, Barlow drafted the "Declaration of the Independence of Cyberspace" and posted it to the Internet. According to Barlow the "Governments of the Industrial World" had become "weary giants of flesh and steel."[8]

In the "Declaration of the Independence of Cyberspace" quoted by Turner, Barlow directly addressed the leaders of the world as follows:

> Your legal concepts of property, expression, identity, movement, and context do not apply to us. They are based on matter. There is no matter here. Our identities have no bodies, so, unlike you,

we cannot obtain order by physical coercion. We believe that from ethics, enlightened self-interest, and the commonwealth our governance will emerge.[9]

In Mario Savio's speech and in Barlow's declaration, we find the basic concepts of the cyberculture as ideology of the general intellect, and also the essential misconceptions leading cyberculture into the traps of neoliberal dogma.

In his declaration, in fact, Barlow rightly reclaimed the radical novelty of immaterial production, and the incompatibility of the new technological world with the old legal system of property and privatization. But he was simultaneously deadly wrong when he wrote that "our identities have no body," because although the body can be denied, and forgotten, it is always pulsating behind the screen. The virtualization of language changes the bodily conditions of life and of communication, but does not eliminate bodily existence.

The denial of the bodily and psychic effects of virtualization was the fundamental fallacy of the cyber-ideology that flourished in the 1990s. This fallacy led to the creation of the sweetish cyber-utopia that came together with neoliberal politics.

In his book, Fred Turner retraced the trajectory ranging from the creation of the computing metaphor by Norbert Wiener, to the intellectual experience of the *Whole Earth Catalogue* and the "Whole Earth 'Lectronic Link" founded by Steward Brand, and contemporary of the technical construction of the electronic networks in the 1980s, until the elaboration of a sort of global mind theology based on the idea of the networked-economy's invisible hand.

Wiener began to imagine duplicating the human brain with electrical circuits. By 1948, he had transformed the computational metaphor into the basis of a new discipline. In his books, he defined cybernetics as a field focused on the study of messages as a means of controlling machinery and society, with machinery seeming to include, by analogy, biological organisms. [...] Wiener believed that biological, mechanical, and information systems, including then emerging digital computers, could be seen as analogues of one another. All controlled themselves by sending and receiving messages, and metaphorically at least, all were simply patterns of ordered information to a world otherwise tending to entropy and noise.[10]

Developing the implications of his own technical and theoretical realizations, Norbert Wiener pointed out that cybernetics might lead to a malevolent automation of human behavior. As Wiener himself suggested, computers might step beyond the reaches of human control and begin to act on their own. Over the next fifteen years, interestingly, Wiener sought out union leaders in order to find out how workers might combat the threats posed by the advanced effects of computing. The history of high technological research and development is marked by this kind of paradoxical attitude on behalf of engineers and scientists.

From the moment a group of scientists was summoned by the American army and invited to work in the framework of the Manhattan Project to produce the nuclear bomb, the terrible alternative contained in scientific work became evident.

Modern history has been marked by the interaction, the conflict, the negotiation, and the alliance between the intellectual, the merchant and the warrior.

The intellectual is the heir of human labor, the bearer of the intelligence accumulated by the infinite succession of acts of labor and the infinite series of acts to refuse labor. The refusal of labor induces the evolutionary motion of intelligence. Intelligence is the refusal of work that is turned into socially useful form. Because of intelligence, it becomes possible to replace human labor with machines. Because of the refusal of work, science is pushed forward, developed, put into practice. Since the outset, modern science has been aware of its function in this respect.

Knowledge multiplies the human capacity to produce useful things and enhances freedom for all humans by reducing the necessary labor time to produce what society needs. This means that knowledge is potency. The merchant and the warrior want to turn knowledge—this concrete useful potency—into an instrument of power, the abstract power of money, or the destructive power of violence. And to this end, they have to subdue the intellectual, the sage, and the scientist. But this does not occur easily, because knowledge does not tolerate domination. Thus the warrior and the merchant resort to traps and deceit in order to subject the potency of thinking to the power of money and violence.

In a 1958 book entitled *Brighter than a Thousand Suns*, Robert Jungk tells how, during the Second World War, the warrior captured the sage through the history of the Manhattan project that led to the realization of the nuclear bomb.[11] A group of scientists faced blackmail: "Hitler might be preparing a nuclear bomb. We need to hurry, in order to anticipate him."

The US government managed to convince a group of scientists to surrender to this blackmail. The effect of the intellectual's surrender to the warrior was Hiroshima. This was the beginning of the struggle for the liberation of the intellectual from the warrior, a struggle that culminated in 1968. 1968 represented, foremost, the intellectual's refusal to lend her or his knowledge to the warrior, as well as the decision to put knowledge at the service of society. But then the merchant came along to seduce the intellectual and to subdue her or his knowledge to the domination of technical-economic automatisms.

The evaluation of truth in the field of knowledge was thus subjected to the criteria of competitiveness, economic efficiency, and the pursuit of maximum profit. In the decades inaugurated by Thatcher and Reagan, knowledge was put to work in conditions of absolute dependence on capital. Science had been incorporated into the automatisms of technology, deprived of the possibility of changing the finalities that guided its functional operation. The intensive application of knowledge to production led to the creation of the digital technosphere, which emanated effects of extraordinary power. But this power was subject to the technical automatisms in which power is embodied. Constrained within the categories of the profit economy, technology increased the productivity of labor, whilst simultaneously multiplying misery, subordination of human beings to wage labor, precariousness, unemployment, and a new sort of alienation that generated unhappiness.

Cognitarians

With the Seattle riots against the World Trade Organization summit in 1999, a movement emerged aiming at the social, epistemic, and technological re-composition of cognitive labor. This implied that scientific research should be autonomous from the interests of the merchants.

The diffusion of open source practices, both in information technology and in biotechnology, that is, free access to the products of intellectual innovation and media activism, have been manifestations of this struggle for the autonomy and self-organization of cognitive work.

In the years marked by mobilization against the institutions of global governance such as the WTO, the FMI, and the G8, among others, cognitive workers took the lead in a wide movement erroneously labeled *anti-globalization*. In fact, this was the first global movement, and it was directed against capitalist globalization, not against globalization itself.

Because of the opening of the stock market and mass Net-trading in the 1990's, wide participation in capital profit was possible. This resulted in the dotcom economy. Cognitive laborers invested their competence, knowledge, and creativity, and found ways to create enterprises in the stock market. For a few years, the creation of virtual enterprises was the zone of encounter between financial capital and cognitive work. A new form of self-enterprise glorified both the autonomy of labor, and dependence on the market. But after a decade of growth, the alliance between cognitive labor and recombining capital broke down in the spring of 2000. The stock market crash of April 2000 was the beginning of a political crisis in the relation between capital and cognitive labor. Many different factors provoked this rupture. The first was the collapse of the psychic and social energies of cognitive labor. Overexploitation, the acceleration of life rhythms, the 24-hour working day of mobile precarious workers, depression, and the excessive use of stimulant drugs to sustain the pace of excessive labor provoked the financial crash of the Net economy, and consequently opened the way to the mass precarization of cognitive labor.

The process of self-organization of the general intellect that was implied in the dotcom experience and in the process of the shared creation of the Internet was sapped and overthrown by the coercive privatization of the products of collective knowledge and by a process of definancing and privatizing public educational institutions. A dismantling of sorts of the general intellect has been underway since the beginning of the new century. The Bush wars restored the primacy of the old military economy, subjecting the new technology to old military systems. This has led to the submission of the general intellect.

The dotcoms were a laboratory for the formation of a new model of production and of markets, but in the end the market was suffocated by monopolies, and the army of self-entrepreneurs and venture micro-traders was dispersed and finally subjected to precarious forms of employment. Corporations gained the upper hand within the cycle of the Net-economy, and allied themselves with the dominant group from the old economy, blocking and perverting the project of globalization itself. Neoliberalism produced its own negation: monopoly domination and state-military dictatorship. Cognitive workers were marginalized and reduced to the precarious condition of salaried submission. The promise implicit in the ideology of the virtual new economy was one of high compensation, and participation in the economic fortunes of the system. But since the collapse of the virtual economy in the year 2000, cognitive work has been subject to precarization.

The intellectual has become a cognitarian, that is, simultaneously a cognitive worker and a proletarian.

The Artist, the Engineer, and the Economist

The intellectual, the merchant, and the warrior were the dominant characters of the fable that we call modernity. But the intellectual function is far from consistent and undifferentiated. On the contrary, the intellectual function is penetrated by an internal conflict, whose dynamics I'll try here to outline. In the sphere of cognitive work, the artist, the engineer, and the economist have been the main characters of the fable that we call the general intellect. Their game was at the core of the intellectual dynamics of intellectual life of the past century. But who won?

The artist, by which I mean both the poet and the scientist, embodies the excess of knowledge and of language, the excess that produces the rupture of their established frames. The artist is the creator of new concepts and of new percepts, disclosing new possible horizons of social experience. The artist speaks the language of conjunction. In artistic creation, the relation between sign and meaning is not conventionally fixed but pragmatically displaced and constantly renegotiated.

The engineer is the master of technology, the intellectual who transforms concepts into projects, and projects into algorithms. The engineer speaks the language of connection. The relation between sign and meaning is conventionally inscribed in engineering. The engineer is a producer of machines, technical combinations of concepts, algorithms, and physical matter performing in accordance with concepts. Contemporary cognitive work is predominantly the work of engineers.

The third figure of the contemporary general intellect is the economist, who is the fake scientist and the real technologist who has been charged to reduce the combined power of the artist and the engineer into the established rules of capitalist accumulation.

Economists are more priests than scientists. Their discourse aims to subject the activity of others to the rule of economic expansion. They denounce society's bad behavior, urging people to repent for their debts, threatening to inflict inflation and misery for people's sins, and worshipping the dogmas of growth and competition.

Scientific methodology is different from the methodology of the economist.

What is science after all? I would simply say that science is a form of knowledge that is free from dogma, and that aims to extrapolate general laws from the observation of empirical phenomena, drawing from this extrapolation the ability to predict something about what will happen next. But science is also able to transcend any kind of causal determinism, and to understand the types of changes that Thomas Kuhn labeled paradigm shifts. This means that scientific innovation is essentially the transgression of the established limits of knowledge.

As far as I know, economics does not correspond to this description. First of all, economists are obsessed with dogmatic notions such as growth, competition, and gross national product. They profess that social reality is in crisis if it is does not conform to the dictates of these notions. Secondly, economists are incapable of inferring laws from the observation of reality, instead preferring that reality harmonize with their own presuppositions. As a consequence, they cannot predict anything—and experience has often shown the economist's inability to predict change and contingencies. Finally, economists cannot recognize changes in the social paradigm, and they refuse to adjust their conceptual framework accordingly. They insist instead that reality must be changed to correspond to their outdated criteria.

Physics, chemistry, biology, astronomy conceptualize a specific field of reality, while in the schools of economics and in business schools the subject of teaching and learning is a technology, a set of tools, procedures, and pragmatic protocols intended to twist social reality to serve practical purposes: profits, growth, accumulation, power. Economic reality does not exist. It is the result of a process of technical modeling, of submission, and exploitation.

The theoretical discourse that supports this economic technology can be defined as ideology, in the sense proposed by Marx—who was not an economist, but a critic of political economy. Ideology is in fact a theoretical technology aimed at advancing special political and social goals. And economic ideology, like all technologies, is not self-reflexive and therefore cannot develop a theoretical self-understanding. It cannot reframe itself in relation to a paradigm shift.

The economist is the entangler of the engineer. Engineering as a technology can be linked to art and science, and transforms conceptual creation into technical *dispositives* for the organization of social life. But over the course of the past century, engineering has been finally subjected to economics, through a reduction of the technical possibilities of machines to single-minded economic determination.

When the engineer is linked to the artist, he produces machines for the liberation of time from work and for maximum social usefulness. When the engineer is controlled by the economist, he produces machines for the entanglement of human time and intelligence with the iteration of the maximization of profit, and the accumulation of capital.

When the engineer is linked to the artist, his horizon is the infinity of nature and language. When he is controlled by the economist, his horizon is economic growth, and for the sake of this dogma he destroys nature, and language has to be compatible with the code.

Capitalism today can no longer semiotize and organize the social potency of cognitive productivity, since economic conceptualization is too narrow for the intellectual potency of society, and since intellectual potency demands a trans-economic dimension.

The shift from the industrial to the semiotic form of production has propelled capitalism out of itself, out of its ideological self-conception. Economists are dazzled by this transformation.

In the present crisis of capitalist globalization, the problem is the following: Is it possible to disentangle activity and knowledge from the grip of the economic, semiotic paradigm? Has the economist totally subjugated the engineer, who previously captured the artist, or can the engineer be freed from economic limitations, and reframe technology according to the intuitions of science and sensibility?

Virtual Utopia

The future is the imaginary space that utopia tries to colonize. This colonization never perfectly succeeds, although it always leaves a few traces in the body of the future.

The twentieth century was colonized by the utopia of futurism, which was based on the rhetorical celebration of the acceleration and violence that actually pervaded the social and aesthetic reality of the century, but that also exhausted its source of energy.

The futurist colonization of the world has exhausted the future itself. For the future is not merely a dimension in time, but a cultural projection, the internalization of time as expansion. In this sense, the future comes to an end when expansion is over, when we become physically, aesthetically and conceptually aware of the exhaustibility, and the actual exhaustion, of energy.

The trajectory of late modernity started with futurism and ended with punk, when culture became aware of the incipient exhaustion of modern energy.

We can reframe this in terms of the notion of the frontier. The future is the displacement of the frontier, the expansion of mapped and colonized territory. When the process of the physical colonization of the planet reached its peak, when the physical territory of the world was completely colonized, when every form of activity was finally subjected to the rule of the market, and when every form of language was subjected to economic reason, then the word *frontier* became an empty word. A new dimension was needed for the unceasing process of deterritorialization that is the engine of capitalism.

Then, after the fall of the Berlin wall, during the decade of globalization, a new frontier surfaced, a new future was imagined, and a new utopia conceived and propagated: the cyber-future, beyond the virtual frontier. The cyber-utopia of the 1990s displaced the frontier from the physical to the virtual dimension.

Since the 1960s, the utopian background of cyberspace was nurtured by freaks, libertarians, and poets, and in fact, the creation of the Internet in the 1990s cannot be separated from the utopian flourishing of psychedelic imagination and of libertarian politics. The World Wide Web has carried a mystical utopian charge for many, ever since the time to which Marshall McLuhan was referring to in the following:

> Today we have extended our central nervous system itself in a global embrace, abolishing both space and time as far as our planet is concerned.[12]

In the first issue of *Wired*, the magazine that has promoted the network as a force of total renovation since 1993, editor-in-chief Louis Rossetto wrote:

> The digital revolution is whipping through our lives like a Bengali typhoon bringing with it social changes so profound their only parallel is probably the discovery of fire. [...] *Wired* is about the most powerful people on the planet today—the digital generation. These are the people who not only foresaw how the merger of computers, telecommunications, and the media is transforming life at the cusp of the new millennium, they are making it happen.

In the 1990s, the cyber-utopia turned into the promise of boundless economic expansion, and also a promise of peace and cooperation. Peter Schwartz and Peter Leyden wrote a text about the new economy fostered by networking, entitling it *The Long Boom*. After Clinton's victory in the 1992 elections, the cyber-futurist utopia fueled expectations of a long age of economic expansion, and of melting with the market theology that is at the core of neoliberal fundamentalism.

Esther Dyson, George Gilder, Alvin Toffler, and George Keyworth wrote a "Magna Carta for the Knowledge Age," declaring in their preamble:

> The central event of the 20th century is the overthrow of matter. [...] The powers of mind are everywhere ascendant over the brute force of things.[13]

The emancipation of the mind from matter was the simultaneous effect of the deployment of knowledge technology with the comeback of the old psychedelic dream.

"After all, what had LSD users hoped to accomplish if not the overthrow of matter by the powers of mind?," noted Fred Turner in *From Counterculture to Cyberculture*.

The convergence of cyber-utopia and neoliberal thinking was clearly articulated by Kevin Kelly in the book *Out of Control*. For Kelly,

> The Net is an emblem of multiples. Out of it comes swarm being—distributed being—spreading the self over the entire web so that no part can say: "I am the I." It is irredeemably social, unabashedly of many minds. It conveys the logic both of Computer and Nature, which in turn convey a power beyond understanding.
>
> Hidden in the Net is the mystery of the Invisible Hand—control without authority.[14]

Since Adam Smith, the invisible hand has been a crucial concept for market-oriented philosophy. This concept implies the assumption of a perfectly functioning hive-mind, where individual minds work homogeneously according to the same principle, the economic principle.

> The tiny bees in my hive are more or less unaware of their colony. By definition, their collective hive mind must transcend their small bee minds. As we wire ourselves up into a hivish network, many things will emerge that we, as mere neurons in the network, don't expect, don't understand, can't control, or don't even perceive.[15]

The network opened a new horizon both to human communication and to capitalist expansion—but the perfect imagination of the Net as invisible hand only worked up to a certain point.

In the year 2000, just a few years after the dawn of the cyber-utopian world, the celebration was suddenly over, and the dotcom crash provoked an economic crisis and opened the gate to a dramatic change. At the turn of the century, the virtual bubble burst, unleashing the sudden awakening of ghosts of aggressiveness and war. Precarization of cognitive work and massive unemployment became the defining features of the labor market, while the euphoria of the 1990s gave way to pervasive gloom.

Why so? I think that the crucial flaw of cyber-utopia was the assumption that "matter [had] been overthrown," and that, as Barlow put, "our identities have no bodies."

This assumption was based on a double mistake. First of all, the sphere of electronics is not immaterial; it simply pertains to a different dimension of matter than the mechanical.

Secondly, and most interestingly, the mind should not be identified as immaterial. According to anthropological, psychological, and neurological points of view, mind must be understood in terms of matter.

Conceptually, only the semiotic sphere can be opposed to matter, and it was the electronic transformation that made it possible to transform social production in a semiotic sense. Semiocapital is the sphere in which economic exchange is predominantly an exchange of signs, but the forces that produce semiotic immateriality are not immaterial.

Cyber utopia forgot, and downplayed, the effects of virtualization on the human psyche and on social subjectivity, because the technophile environment of cyber-utopianism understood the psychological effects of virtualization in an absolutely superficial way.

The 2000 dotcom crash not only exposed the economic frailty of the new economy, but also unmasked the frailty of the networked nervous system. Cyber utopians and neoliberals were not interested in the effects of technology on social subjectivity, since they thought that psychological reality could be reduced to the cognizant surface of consciousness, and totally downplayed the complexity of the unconscious. But this neglect marks the failure of both cyber-utopianism and neoliberal politics. In order to understand the failure of the market theology generated by the cyber-utopia of the 1990s, we must now investigate the unconsciousness of the swarm-mind.

The Soul at Work

Semiocapitalism is based on the exploitation of the soul as a produc-tive force and as a market place. But the soul is much more unpredictable than the muscular workforce which was at work in the assembly line. In the years of the Prozac economy, the soul was happy to be exploited, but this could not last forever. Soul troubles first appeared in the last year of the dotcom decade, when the techno-apocalypse was announced under the name of the millennium bug. The social imagination was so charged with apocalyptic expectations that the myth of the global techno-crash sent a convulsive wave through the world. In the end, nothing happened during the millen-nium night, but the global psyche teetered on the brink of an abyss.

What is essential about the social transformations caused by the digitalization of production is not the loss of regularity of the labor relation, which, after all, has always been precarious notwithstanding legal regulations, but rather the dissolution of the person as an active, productive agent, as labor force.

The cyberspace of global production should be seen as an immense expanse of depersonalized human time. Info-labor, the provision of time for the elaboration and the recombination of segments of info-commodities, is the point of culmination of the process of abstraction from concrete activities that Marx analyzed as a tendency inscribed in the capital-labor relation.

The process of abstraction of labor has progressively stripped labor time of every concrete and individual particularity. The atom of time of which Marx speaks is the minimal unit of productive labor. But in industrial production, abstract labor time was impersonated by a physical and juridical bearer, embodied in a worker made of flesh and bone, with a certified and political identity. Naturally, capital did not purchase a personal disposition, but the abstract time of which the workers were bearers. The capitalist was forced to hire human beings, and therefore to deal with their physical weaknesses, sicknesses, human rights, and face trade unions as well as the political demands these subjects bore.

When we move into the sphere of info-labor, there is no longer a need to buy the availability of a person for eight hours a day indefinitely. Capital no longer recruits people, but buys packets of time, separated from their interchangeable and occasional bearers. In the Net economy, flexibility has evolved into a form of fractalization of work.

Fractalization means the modular and recombinant fragmentation of the time of activity. The worker no longer exists as a person. He or she is only an interchangeable producer of micro-fragments of recombinant semiosis that enter into the continuous flux of the Net.

Capital no longer pays for a worker to be available to be exploited for a long period of time; it no longer pays a salary that covers the entire range of the economic needs of the person who works.

The worker, a machine endowed with a brain that can be used for fragments of time, is paid for his or her occasional, temporary services. Work time is fragmented and cellularized. Cells of time are for sale on the Net, and businesses can buy as much as they want without being obligated in any way concerning the social protection of the worker. Depersonalized time has become the real agent of the process of valorization, and depersonalized time has no rights, and no needs. It can only be either available or unavailable.

The language machine recombines and connects the time fragments necessary to produce info-goods. The human machine is there, pulsating and available, like a brain-sprawl in waiting. The extension of time is meticulously cellularized: cells of productive time can be mobilized in punctual, casual, and fragmentary forms. The recombination of these fragments is automatically realized in the network. The mobile phone is the tool that enables the connection between the needs of semio-capital and the mobilization of the living labor of cyberspace.

The ringtone of the mobile phone calls workers to reconnect their abstract time to the reticular flux.

In this new dimension of labor, people do not have any right over the time that they formally own. That time does not really belong to them, because it is separated from the social existence of those who make it available to the recombinative cyber-productive circuit. The time of work is fractalized, reduced to minimal fragments that can be reassembled, and this fractalization enables capital to constantly find the conditions for minimal salaries. Fractalized work can punctually rebel in some knot of the net, but this does not set into motion any wave of struggle.

For struggles to form a cycle, laboring bodies must be in spatial proximity, and in an existential, temporal continuity. Without such proximity and continuity, cellularized bodies lack the conditions to experience the kind of affectivity that enables social solidarity.

Behavior can only become a wave when there is continuous proximity in time, which info-labor no longer allows.

Cognitive activity has always been the basis of all human production, even that of a more mechanical type. There is no process of human labor that does not imply an exercise of intelligence. But today, cognitive capacity is the essential productive resource. In the sphere of industrial labor, the mind was put to work as a repetitive automatism, the physiological support of muscular movement. Today, the mind is at work in many varying ways, because language and relations are continuously changing. The subsumption of the mind in the process of capitalist valorization therefore leads to a true mutation. The conscious and sensitive organism is subjected to competitive pressure, to an acceleration of stimuli, and to constant attentive stress. As a consequence, the infosphere in which the mind is formed and enters into relations with other minds becomes a psychopathogenic atmosphere. To understand semiocapital's infinite game of mirrors, we must outline a new disciplinary field, delimited by three aspects: the critique of the political economy of connective intelligence; the semiology of linguistic-economic flows; and the psychochemistry of the infosphere.

6

The Swarm Effect

Absolute Capitalism

In my opinion, it is inaccurate to define the contemporary economic system as neoliberal capitalism. Neoliberalism is only the ideological justification of the transformation that happened in the last decades of the twentieth century. It is also inaccurate to define it as monetarism, because the role played by the variation in money supply is simply a technical aspect.

I think that this contemporary transformation should be appreciated from the point of view of long-term human evolution as the turning point beyond the age of humanism.

Modern bourgeois capitalism was a product of the humanist revolution, and the bourgeoisie was the class that embodied the values of a humanist freedom from theological destiny. But the bourgeois characterization of the economic system dissolved as an effect of the end of the preponderance of industry in capital accumulation, and as an effect of the deterritorialization of the production process.

Semiocapitalism took the place of industrial capitalism—because the production and exchange of abstract signs came to play a predominant role in the overall process of accumulation.

Financial abstraction is the extreme manifestation of this predominance.

The expression *cognitive capitalism* also seems incorrect to me. It is not capital that can be labeled cognitive, but labor. The capitalist is not the subject of cognitive activity, but rather the exploiter. The bearer of knowledge, creativity and skills, is the worker.

The term semiocapitalism seems to me to provide a suitable definition of the present economic system at the global level.

However, in order to grasp the political dimension of the transformation that neoliberal deregulation has brought about, I think that we should refer to it as capitalist absolutism.

The bourgeoisie fought a battle against early modern absolutism, after taking advantage of the effects of national unification and social regulation that absolutist monarchs had enforced on traditional social life forms. The bourgeois fight against monarchic absolutism was part of the battle for the liberation of privately owned enterprises from the control of the state, and also (and most interestingly from my point of view at present) a battle for the rule of law, for the constitutional limitation of the actions of the monarch.

The term *absolutism* comes from the Latin *absolutus*, which can be translated as "emancipated from any limitation." In this context the word *absolute* means "not limited by restrictions, unconditional, unconstrained by constitutional or other provisions."

The bourgeoisie reclaimed the rule of law in order to limit the power of the feudal aristocratic class and of the monarchs. It then accepted that the rule of law be limited by its own economic expansion, because the bourgeois was concerned about the community of workers and consumers, about the territory of the city, and about the future of the common welfare which was obviously linked to the future of his own investments.

This is why the bourgeois class accepted the democratic pact, and consented to negotiate with the working class. The bourgeoisie could not be indifferent towards the destiny of its territory, and towards the community of those who worked there. Workers and bourgeoisie shared the same urban space, and the same future. If the economy crumbled, it was also a misfortune for owners, albeit it was a much worse one for workers and their families.

Following the rise of financial capitalism, the deterritorialization of production and exchange, and finally the emergence of a virtual class that was not identifiable in territorial terms, a general process of deregulation ensued.

First of all, the fact that corporations were globalized, with global characteristics and a global sphere of activity, hindered and rendered impossible any legal control of their activity.

The sovereignty of nation-states lost its effectiveness, and global corporations gained absolute freedom, no longer responding to local authority and shifting their immaterial assets from one location to another. Secondly, the globalization of the labor market destroyed the unionized power of workers, and opened the way to a general reduction of salaries, to an increase in exploitation, and to the removal of any regulations concerning work conditions and working time. The plague of child labor has returned in many areas of the world, while retirement time has been postponed and people are forced to work until the age of 65, 67, or even later.

Furthermore, limits to the exploitation of physical resources and to environmental pollution are systematically ignored by corporations whose only parameter is the growth of profit.

I think that our contemporary global reality should be defined as absolute capitalism. The accumulation of value, increase in profits, and economic competition are the only effective regulations of this world, its all-encompassing priorities, and its almighty power. All other priorities or interests, including the survival of the planet and the destiny of the generation to come, are of no import.

Compared to the history of bourgeois industrial capitalism, the relation between social welfare and financial profit has been inverted. In an industrial economy, profits increased when citizens had enough money to buy goods produced in factories. In the sphere of financial capitalism, stock market indexes go up when salaries fall and social welfare crumbles.

It is not surprising that the few hundred billionaires listed in Forbes Magazine have hugely increased their capital during recent years that have been marked by rising unemployment and misery, and public spending cuts.

Is the financial class getting richer despite the general impoverishment? No, the financial class is getting immensely richer precisely *because of* the general impoverishment of society, and thanks to it.

This is why I think that capitalist absolutism is the most-suited definition for the present system.

The Matrix and the Cloud, Again

Absolute capitalism is destroying the protections that were created by modern social civilization, and we are returning to the original, naked condition of human existence: precariousness. The truth about human existence—which modernity tried to rationalize and to provide protection from—is brutally revealed. The rule of law has been cancelled by decades of neoliberal globalization, and welfare structures are currently destroyed by absolute capitalism.

Deterritorialized fragments of precarious time are now scattered throughout the physical world, fragments of life that are unable to meet and conjoin, but that are perfectly able to interact when they are recombined by the digital net. Metropolitan sprawl is filled by people who meet for an instant, and then part and never meet again. They connect, in the act of production, in the abstract process of syntactic exchange, and then disappear in the chaotic meltdown of precarious deterritorialization. The conditions of social solidarity, togetherness, long-lasting collaboration in the same place of work, and urban proximity, have been dissolved.

The old fashioned order of political reason is lost, as is the regular relation between time and value. All that is solid melts into air, as Marx wrote in 1848, and, no longer regulated by shared conventions, social existence is turned into a cloud, water vapor endlessly moving, condensing, and dissolving, changing form and degrees of density.

It has become impossible to draw a map of urban pathways, or a map of the labor market, as it has become impossible to foretell where the molecules of social life will migrate. How can one decide in a rational way about the future? One cannot. This is precariousness: the unmappable cloud.

Simultaneously, however, the network of capitalist accumulation is permeating both the cloud and the Matrix. Fragmented life is re-coded, fractalized, and recombined through the rules of connection that are embedded in the technical interfaces of social communication. The Matrix connects the neuronal pathways of scattered bodies, distant in time and space, but that connect in the operational framework of the code. The Matrix is therefore subjecting collective cognition to the regular patterns of the algorithm, and is turning the cloud into a swarm.

The European Cult of Governance

What happens to social solidarity and empathy when social space is invaded by precariousness? What happens to political freedom and democracy in the systemic order of governance?

In order to answer these questions, let's go back to the contemporary European crisis. The history of the European crisis, which is in full swing as I'm writing these pages, is evidence of the loss of effectiveness of voluntary action and of democratic decision-making. In fact, the financial system, which has taken the upper hand in the political dynamics of the European Union, has inserted a chain of automatic implications into social economy, depriving democracy of any content, at least in the social field.

When the elected prime minister of Greece, Georgios Papandreou, declared the intention to call for a referendum about applying the austerity measures imposed by the European Central Bank, he was obliged to resign overnight. After his dismissal, a former consultant of a global financial agency was summoned and installed in his place.

When Italians voted massively against austerity during the February 2013 elections, the president of the European Central Bank, Mario Draghi, reacted by saying ironically that the outcome of the election should not be overstated, since Italian economic policy was on autopilot. In other words, it was governed by the automatisms implied in the *Fiscal Compact*, a set of financial measures imperatively imposed on the national parliaments of the continent.

The assumption that human beings have the possibility of governing their own history has never been truly well founded. Human will has always been a factor among many others in the determination of the historical process. But in the modern age— since Machiavelli to be precise—that factor has gathered momentum thanks to scientific knowledge, and to the establishment of the political space of bourgeois civilization. Government, as a rational project leading to the controlled and planned transformation of social reality, had been possible as long as the bourgeoisie had been able to subdue the complexity of human passions to the interests of economic growth, and to the overall improvement of social life. Both in the authoritarian forms of monarchic absolutism, and in the liberal forms of democracy, the word *government* meant the ability of the brain to supervise the development of social forces. Knowledge, anticipation, planning, but also violence were the tools for governing society, and for reducing it to this common goal, as long as this reduction was possible.

With the expansion of economic power and the overly-fast acceleration of informational exchange, government has turned into an empty word, and human will has given way to the automatic self-replication of information and power.

Although the illusion of democracy continues, and the rhetoric of democracy has been promoted as the universal myth of political discourse, democracy is nothing but a ritual when it comes to the formation of power in the sphere of the economy. The European Union, which was built on the promise of enhancing the space of democracy, has recently exposed the emptiness of democratic discourse in the digital age.

Governance is the keyword of the European construction, but it is also the keyword of financial capitalism worldwide. Governance is the replacement of human will and democratic government with automatic systems that have logical and technological implications.

It is pure functionality without meaning, the automation of thought and will.

In the concept of governance, abstract connections between living organisms are embedded in language and in the interactions of social life. The concatenation of choices, decisions, and priorities is identified with logical rationality and objectified into technical protocols.

At its beginning, the European entity was conceived as the possibility of overcoming passions, nationalist, ideological, or cultural passion, all dangerous marks of belonging. Even the aesthetics of Europe are marked by an intentional frigidity which may be read as an attempt to distance it from the romantic imprinting of European modernity which led to the catastrophic wars of the twentieth century.

From this point of view, the European Union is a perfectly postmodern construction in which power is embodied by techno-linguistic devices of interconnection.

Through its attempt to fuse and overcome strong national identities, the French and German identities above all, the European Union has no cultural identity. Because of this, Europe has founded its identity on prosperity. For several decades, prosperity has been the unifying mark of a heterogeneous Union. The European entity has not identified with political passions, with great ideological visions, or with charismatic leaders, but with the cool image of bankers.

This identification worked until the end of the first decade of the new century. As long as financial capitalism guaranteed a growing level of prosperity, as long as the monetarist rule helped the economy to grow, Europe thrived.

But what next? What if Europe loses its status of prosperity and growth?

As I'm writing these lines, in the year 2013, a question is being anxiously asked: Will Europe survive the financial collapse and the upheaval that will follow, as financial architecture has been its only support?

The EU is not a democracy: it is ruled by an autocratic organism, the European Central Bank, and by a financial class that does not respond to citizens or to parliament.

Because of the Maastricht Treatise turned into a system of absolute dogmas, the living societies of European countries are being subjected to a strict neoliberal rule, the cutting of labor cost and the downsizing of the public sphere.

Governance has replaced political will with a system of automatic technicalities that force reality into an unquestionable, logical framework. Financial stability, competitiveness, reduction in labor costs, and increase in productivity: the systemic architecture of the EU is based on these dogmatic foundations, which cannot be challenged or discussed because they are embedded in the technical functioning of technical sub-systems of management. No enunciation or action is operational if it does not comply with the embedded rules of neoliberal governance.

Governance is the management of a system that is too complex to be governed. The concept of government implies the possibility of understanding social processes and cultural expectations, and the ability of the human will, whether despotic or democratic, to control information flows so that a relevant part of the social whole can be controlled and managed. Government is possible as far as the degree of complexity of social information is low. But information complexity has been growing throughout the late modern age, and has finally exploded in the age of the digital networks.

Because of the proliferation of information exchange, the intensity and the speed of circulation of social information has grown too fast for centralized knowledge and for political control. Rational government is therefore impossible, since any critical discrimination and determination of the sequence of events and information is impossible. Here, the governance mode is ushered in. The abstract concatenation of technical functions replaces conscious elaboration, social negotiation, and democratic decision.

The automatic connection of a-signifying segments replaces the dialogic elaboration of an order, and adaptation replaces consensus.

In the sphere of governance, the agencies that enter the social game have to be formatted and made compatible before they start to exchange. Conflicting interests and projects have to bend to the unquestionable rationale of the algorithms that define governance pattern. The rhetoric of system complexity replaces the rhetoric of historical dialectics.

Disruptions have to be managed according to a shared pattern of compatibility.

Complexity, Chaos, and Meaning

In *Out of Control*, Kevin Kelly explains that "the world of our own making has become so complicated that we must turn to the world of the born to understand how to manage it."[1]

From an epistemological point of view, the notion of complexity is nothing but a truism.

In a text published in 1985, Edgar Morin himself wrote that about complexity that it appears as "difficulty and uncertainty, not as precise answer."

The notion of complexity acquires meaningful relevance only if one thinks in terms of information. If we consider the environment as a source of info-stimuli, and the mind as a receiver-decoder, complexity can be defined as a function of the relation between the intensity of the information flow and the pace of the mental elaboration of input.

Chaos surfaces in the concatenation between the mind and the environment, when the information flow is too fast for conscious elaboration.

The word *chaos* thus stands for an environment that is too complex to be decoded by the explanatory grills available to us, an environment in which flows are too quick for conscious elaboration and rational decision. The word *chaos* denotes a degree of complexity that is too dense, too intense, and too fast for our brains to decipher. From this point of view, complexity is considered a measure of the speed of the receiver (the mind) in relation to the speed of the transmitter (the surrounding infosphere).

Sometimes, in science and in life, a sequence of events can reach such a level of complexity that a small perturbation will have huge, unpredictable effects. We speak of chaos when this kind of indetermination spreads everywhere.

The process of mathematization of the world, which is the core of the modern scientific methodology, is an act of reduction of the environment to measure (proportioning and reduction to measurement). The Latin root of the word *reason*, in fact, refers to measure (*ratio*). Measurement cannot be accomplished without a reduction that cuts the extension of what is relevant out of the infinite flow of the signs of the world. The problem of relevance is crucial in the passage from chaos to order, and therefore in the process of civilization. One has to discriminate relevant events in order to measure.

Science cannot be productive of knowledge without establishing limits for investigation. What is inside the established limit is called relevant, what is outside those limits is irrelevant. Similarly the political mind cannot decide without posing limits. Only what is relevant from the point of view of knowledge is actually elaborated by the rational mind. Rational government presupposes the extrapolation of relevant information from an infinite flow. What is relevant, what is not? That question implies the epistemic rationalization of the available flow of information.

Modern scientific culture could only keep reality under rational control through the limitation and the exclusion of irrational mythologies and similar other forms of craziness from the space of decision.

Machiavelli distinguished the sphere of fortune from the sphere of will. The prince was the person (male) who subdued fortune (female) to rational political will.

Fortune was chaos hiding in the folds of human experience.

In order for the prince to govern, he must previously have cut out a narrow string of events in the infinity of fortune. The dark infinity of un-reducible chaos lay at the border of established order. Chaos was noise; order was rhythm.

Ordering rhythm enabled the synchronization of fortune and will, reality and reason. But only a tiny part of the sphere of reality could be synchronized with reason, and only a tiny part of fortune could be synchronized with political will. This tiny part was labeled *relevant* by the ruling intellect of order. The pretense of total government was always an illusion, because the entire multiplicity of world events is ungovernable. But this illusion could work and could produce effects when the infosphere was so thin, and the flow of information so slow, that political consciousness could cut out a small space of relevant social events, and try to protect this space, civilized space, from the surrounding ocean of non-governable matter.

This is why the kingdom of civilization is in crisis today. The acceleration of media flows stimulating the collective brain is breaking the frame of the rhythm that we inherited from the modern age.

Chaos resurfaces when the stream of digital information flows too fast for the rhythms of mechanic theory and political will. As the electronic flow invades the screen of our attention, the protecting fence of relevance is broken, because we can no longer discriminate what is relevant from what is not.

I want to focus now on the production of meaning in the framework of the relation between the mind and its environment. Meaning can be defined as a reduction of reality to a finite concatenation of enunciations.

When the infosphere is slow enough to be screened and scanned by the mind, then we can extract meaning from it, and find a common rhythm. This ritornello is the synchronization of mental activity with the environment.

When the infosphere saturates our attention time, and the semiotic flow goes too fast for our mind to process information in a rational way, we speak of complexity.

Within certain conditions of speed—when the info-flow is slow—a rational model of government can control the environment and can decide between alternative possibilities. But when the intensity of information and the speed of the infosphere overrun the mind's pace of elaboration, then the mind can no longer extract meaning from experience and the psychosphere is affected by a sense of confusion. Meaning can no longer be grasped since we cannot extract from the infinite flow a finite explanation as a workable tool for social interaction and understanding. At that point, social order can only be produced by syntactic selectors of meaning and automatic deciders. Semantic interpretation is no longer possible because time is too short. Decisions must be made by default by purely syntactic machines.

Governance is the combination of the automatic decisions made by syntactic machines.

Swarm and Connectivity

A multitude is a plurality of conscious and sensitive beings who do not share a common intentionality, and who do not show any common pattern of behavior. Crowds shuffling in the city move in countless different directions, with incalculable different motivations. Everyone goes his or her way, and the intersection of those movements makes a crowd. Sometimes, the crowd seems to be moving in coordination: people run towards the station because a train is expected to leave soon, or people stop at traffic lights. But everyone moves following his or her will within the constraints of social interdependency. This crowd is a multitude, since it escapes any common intentionality, and common direction.

A network is a plurality of organic and artificial beings, of humans and machines that perform common actions thanks to procedures that make their interconnection and interoperation possible.

If you do not adapt to such procedures, and don't follow the technical rules of the game, you are not playing the game. If you don't react to certain stimuli in a way that complies with the protocol, you don't belong to the network. The behavior of people who are part of a network is not random as are the movements of a crowd, because the network implies and prearranges pathways for the networker.

As I wrote in *The Uprising*, the multitude, the network and the swarm comprise different forms of social organization. The first is plurality of conscious and sensitive beings who share neither a common intentionality nor a pattern of behavior. The second implies rules that must be followed by the networked, who perform common actions thanks to procedures that make their interconnection and interoperation possible. Third, the swarm is a plurality of living beings who follow rules embedded in their neural systems, resorting to common, automatic attributions of meaning, and conforming behavior.

As semiocapital is introducing techno-linguistic machines in the flow of communication, the living body of society is turning into a swarm. As a consequence, the very concept of human freedom is transformed. Forms of dissidence can be expressed, and acts of refusal can be performed, but they will be ineffective because they cannot change the direction of the swarm, nor can they change the way the swarm brain is processing information.

The entomologist Morton Wheeler calls the swarm a superorganism that emerges from the mass of ordinary insect organisms. As Kevin Kelly writes in *Out of Control*:

The hive possesses much that none of its parts possesses. One speck of a honeybee brain operates with a memory of six days;

the hive as a whole operates with a memory of three months, twice as long as the average bee lives.

Kelly, a writer on biology who studies informational social systems, comes to the following conclusion:

> There is a sense in which a global mind also emerges in a network culture. The global mind is the union of computer and nature—of telephones and human brains and more. It is a very large complexity of indeterminate shape governed by an invisible hand of its own. We humans will be unconscious of what the global mind ponders. This is not because we are not smart enough, but because the design of a mind does not allow the parts to understand the whole. The particular thoughts of the global mind—and its subsequent actions—will be out of our control and beyond our understanding.[2]

This is a description of the process that has developed during recent decades, as social systems, incorporating info-machines, and bio-machines have become too complex for human intelligence to understand, and for human will to govern. The troubling side of this process is that humans can no longer stop the machinery they have created, and can no longer correct embedded choices.

Inside a network, human language can only be operational when it obeys the embedded rules of syntactic order and of semantic compatibility. Linguistic acts that do not obey the rules of code-compliance are simply discarded: the bio-informational super-organism reads human language and discards it as noise.

In a text entitled *Networks, Swarms, Multitudes*, the biologist Eugene Thacker studies the analogies and differences between collectivity and connectivity, and underlines that collectivity implies a certain degree of connection, while the contrary is not true, connectivity does not imply the existence of a collective.

Speaking about swarms, Thacker writes:

—A swarm is an organization of multiple, individuated units with some relation to one another. That is, a swarm is a particular kind of collectivity or group phenomenon that may be dependent upon a condition of connectivity.

—A swarm is a collectivity that is defined by relationality. This pertains as much to the level of the individual unit as it does to the overall organization of the swarm. Relation is the rule in swarms.

—A swarm is a dynamic phenomenon (following from its relationality). This differentiates it from the concept of a "network," which has its roots in graph theory and spatial modes of mathematically understanding "things" (or nodes) and "relations" (or edges). A swarm always exists in time and, as such, is always acting, interacting, interrelating, and self-transforming. At some level "living networks" and "swarms" overlap. [...]

Studies in network science, swarm intelligence, and biocomplexity all define self-organization as the emergence of a global pattern from localized interactions. This paradoxical definition makes swarms interesting—politically, technologically, and biologically—for it imputes an intentionality-without-intention, an act-without-actor, and a heterogeneous whole. In swarms there is no central command, no unit or agent that is able to survey, oversee and control the entire swarm. Yet the actions of the swarm are directed, the movement motivated, and the *pattern* has a *purpose*. This is the paradox of swarms.

In fact, the tension within swarms, as both political and biological entities, is a tension between pattern and purpose. Organization does not necessarily imply a reason for its own existence, unless organization itself is the reason. At one pole is a highly directed, purposeful collectivity, such as a crowd of demonstrators (whose purpose may be to block city streets or obtain visibility), or, on purely biological terms, a swarm of army ants (whose purpose is to look for a food resource). Such collectivities may be called swarms, in that they fulfill two basic components of swarms: they exhibit global patterns from local interactions, and they exhibit a directional force with intention that is without centralized control. At another pole are collectivities which are also highly ordered and dynamically organized, but which do not display any overt "purpose" or goal, other than to maintain themselves as such. Examples may include a large crowd at a festival or concert, or on a biological level, flocks of birds and schools of fish. While researchers interpret such examples as driven by an evolutionary necessity (and therefore dictated by the purpose of survival), the kind of teleology this exhibits is remote, indirect, and ultimately relies on the explanatory capacity of evolutionary theory. [...]

There are, then, two axes, two different types of tension, and two sets of concepts. On one axis is the tension between *collectivity* and *connectivity*. While connectivity may be a prerequisite for collectivity, collectivity is not necessarily a prerequisite for connectivity as such. Complications arise when a combination of technological euphoria and new social practices lead to an overoptimistic view of connectivity as immediately implying a collectivity. At the extreme point of technological determinism, political forms such as democracy are rendered inherent in both nature and technology.[3]

Connectivity does not imply collectivity. Collectivity, in fact, is a relation of bodies which share a common analogical understanding, which negotiate continuously about the semantic relevance of their linguistic exchanges, about the meaning of their interactions, in a condition of affective inclusion.

Collectivity happens in conditions of conjunction, while a swarm is a connective body with no conjunction, with no conscious affective collectivity. Conjunction emerges from an unmotivated, logically unnecessary attraction whose purpose is not implied in its pattern. Conjunction is a random concatenation, whose only rule is desire.

Furthermore, conjunction has nothing to do with belonging. While belonging entails a necessary implication and presupposes the fixing of an identity, on the contrary, conjunction does not refer to something embedded and natural. Conjunctive acts do not presuppose any meaning, since meaning is *created* by the acts of conjunction.

In conjunction, knowledge is creation, not recognition.

On the contrary, within connective systems there is no knowledge, but merely syntactic recognition. Connection entails an effect of machine functionality, not of meaningful fusion. In connection, communication implies interfacing and inter-operability. Only segments that were previously made linguistically compatible can interact. The network penetrates the social body, inserting connective segments, and converting the body into a swarm.

While conjunction entails a semantic criterion of interpretation, connection requires a purely syntactic criterion of interpretation. The connective agent (or machine) is required to recognize a sequence and to carry out the operation foreseen by the *general syntax* (or operating system). There is no margin for ambiguity allowed in the exchange of messages.

The translation of semantic possibilities into syntactic binary alternatives is the process that leads to the construction of a digital web.

Collectivity takes shape in the sphere of conjunction, when conscious and sensitive organisms enter into a reciprocal relation of mutual transformation, and continuous questioning and ambiguity.

Connectivity, instead, is the logical implication embedded in the bio-info interfaces of techno-language.

Collapse and Subjectivation

In the modern political landscape, the collapse of a system was considered an opportunity for radical change. Revolution is the term that refers to the subversion and the conscious change of existing social structures. This concept, crucial in the lexicon of modernity, does not properly describe the process of change, as it is based on the illusion of total control of social reality through rational will and linear projects of transformation. Although theoretically inaccurate, however, the concept of revolution has been practically useful to describe those radical processes of conscious and voluntary transformation that marked the history of modern times.

Revolutions were often doomed to give birth to violent and totalitarian systems, and generally did not fulfill their utopian projects. However, they transformed social collapses into radical changes, and paved the way to the shift of political power from one dominant group to another.

Neoliberal deregulation can be seen as the last effective revolution in human history. By joining the technical revolution of the digital network, neoliberal capitalism has turned into a system that is simultaneously flexible and resilient.

Thanks to flexibility, neoliberal capitalism succeeded in repressing the social turbulence of the 1970s, and capturing the technical evolution of the 1980s. Although it celebrated democracy as a universal political value, financial deregulation actually destroyed the very conditions of modern bourgeois democracy, and replaced political decision with a system of financial automatisms.

Far from stable, networked capitalism is a system that continuously borders on collapse. But the final collapse never comes, because of the resiliency of the self-regulation that capitalism borrows from the networked system of semiotic exchange.

The more complex a system becomes, the more it is inclined to disruption. At the same time, the more complex a system grows, the less susceptible it is to voluntary control, and therefore to conscious and intentional change.

In 1917, when the Russian political and military systems were on the brink of collapse, Lenin called for the transformation of imperialist war into revolution. As we know, his call was effective. The Soviet Revolution followed, and social morphogenesis took the form of a communist dictatorship.

The history of the two centuries of modernity is full of such examples. Disruptions culminate into collapses, and collapses give way to revolutions. But nowadays, collapse takes the form of disruption, and no longer gives way to revolution. Instead, it leads to the consolidation of power. Morphostasis follows disruption, so that a process of revolutionary morphogenesis seems to be out of reach.

As I argued before, complexity is a relation between time and information. A system is complex when the density of the infosphere saturates the receptivity of the psychosphere, and the speed of the circulation of information overcomes the human ability to elaborate signs in time. When the infosphere is saturated, and becomes too dense and rapid for conscious elaboration, automatic reducers of complexity begin to operate. A disruption is the effect produced by the irruption of an unpredictable event that interrupts a chain or a flow. In the sphere of connectivity, disruptions tend to proliferate because the infosphere overload makes human actors unable to govern the systemic complexity of social and technological structures.

Disruptions can happen because of the unpredictable interference of nature in the technosphere. Take, for example, the cloud from the Icelandic volcano that blocked European air traffic in March 2010. Or disruptions can occur because of the limits of technological control, such as Chernobyl in 1986, or the massive oil-spill in the Mexican Gulf in the late spring of 2010. Disruptions can also take place through the interference of the social psyche in the automatic flow of information, such as panic effects in the financial circuit.

Over the course of modernity, with the slow circulation of information, and, consequently, the effectiveness of political will and rational government, disruptions were considered triggers of social morphogenesis. During a disruption, political power was expected to weaken, social forces were mobilized, and this situation provided an opportunity for revolution. In conditions of low complexity, political reason was able to change social organization in such a way that a new pattern could emerge.

But in our present condition, when the density and the speed of information are too high for conscious elaboration, disruptions tend to be morphostatic, and to reinforce the patterns that produced the disruption itself.

Why so? Why do systems become more resilient when their complexity grows?

Why does society seem unable to create forms of conscious solidarity that can break the system and begin processes of collective autonomy?

Systems become more resilient when complexity grows, because the more complex a system is, the more knowledge must be concentrated, until it becomes inaccessible.

Commenting upon the many disruptions of the year 2010—the Greek financial collapse, the Icelandic cloud covering in the European skies, and the gigantic oil spill in the Gulf of Mexico—in an article entitled "The Great Consolidation," Ross Douthat, op-ed columnist for *The New York Times*, wrote:

> [...] the economic crisis is producing consolidation rather than revolution, the entrenchment of authority rather than its diffusion, and the concentration of power in the hands of the same elite that presided over the disaster in the first place. [...] The panic of 2008 happened, in part, because the public interest had become too intertwined with private interests for the latter to be allowed to fail. But everything we did to halt the panic, and all the legislation we've passed, has only strengthened the symbiosis. [...] Eighteen months after the financial crisis, the interests of [America's] financiers, CEO's, bureaucrats and politicians are yoked together as never before. [...] This is the perverse logic of meritocracy. Once a system grows sufficiently complex, it doesn't matter how badly our best and brightest foul things up. Every crisis increases their authority, because they seem to be the only ones who understand the system well enough to fix it. But their fixes tend to make the system even more complex and centralized,

and more vulnerable to the next national-security surprise, the next natural disaster, the next economic crisis.[4]

These considerations point essentially to the resiliency of the system, but do not answer the second question: Why does society seem unable to create forms of conscious solidarity that are capable of rupturing the system and starting processes of collective autonomy?

In a situation of high complexity, the social body becomes disconnected from the social brain, and sensibility is disconnected from the intellect; social consciousness is jeopardized, fragmented, such that rage against exploitation turns into frustration and self-contempt. Processes of subjectivation do not find the way to social autonomy. Subjectivation must now come under scrutiny. This is what the third part of this text will address.

PART 3

SUBJECTIVATION

Time is out of joint, wrote Gregory Bateson quoting Hamlet. Out of joint, disjointed. Increasing connectivity and the submission of cognitive activity to digital machines has provoked a disjunction between the mutated pace of the connected mind and the pace of the bodily mind. Consequently, the general intellect has been disjoined from its body.

The problem, here, is not the subject, as a given, a static reality. The problem is subjectivation, the process of the emergence of consciousness and self-reflexivity of the mind, where the mind is not considered in isolation, but in the context of the technological environment, and of social conflict.

Subjectivation is also to be understood as morphogenesis, as the creation of forms.

7

Social Morphogenesis and Neuroplasticity

Out of Joint

In the previous section, I outlined a short history of the general intellect, based on the relation between the language machine and the living brains that cooperate in scientific knowledge and in the technical production of the world. Now, I want to investigate the concept of neuropathology and neuroplasticity, in order to envision the mutation that is underway in the social brain, and to appreciate the possibilities that the present composition of the general intellect brings about.

Since the final years of last century, the symptoms of a sort of dissonance and temporal unbalance are multiplying in the sphere of aesthetic sensibility. The rhythm of life is haunted by a sense of acceleration that fragments living experience and sensory perception itself.

Time is out of joint, disjointed. And the general intellect has been disconnected from its body.

Since we know that the brain is equipped with an incredibly high level of plasticity, we can justifiably wonder if the plastic brain will succeed in finding the way out of the labyrinth, and if the plastic neural system will discover a new, possible conjunction between the world and the mind?

How so?

We must focus on the concept of neuroplasticiy, that is, the ability of the brain to reframe the relation between the rhythm of the receiving mind and the rhythm of the transmitter, a chaotic universe sending signs that are no longer filtered by the grids of a shared semiotic order.

Neuroplasticity: Cognitive Mutation and the Brain

Migraine is a well-known disease (although not exhaustively explained, nor efficiently treated). It is generated by physical processes localizable in the brain, and which can produce psychological effects such as depression.

In his book on migraines, Oliver Sacks underlines the "inseparable unity of psycho-physiological reactions," and refers to chaos theory as a framework to explain the phenomenon of migraines. The complexity of the brain makes it impossible to fully describe the processes that migraines involve, but according to Sacks,

> migraine starts as instability, disturbance, a far-from-equilibrium, unstable state, which sooner or later gravitates into either of two relatively stable positions, that of health or that of illness. [...] McKenzie once called Parkinsonism an organized chaos, and this is equally true of migraine. First there is chaos, then organization, a sick order: it is difficult to know which is worse. The nastiness of the first lies in its uncertainty, its flux: the nastiness of the second in its sense of immutable heavy permanence. Typically treatment is only possible early, before migraine has solidified into immovable fixed forms. The term chaos indeed may be more than a figure of speech here, for the sort of instability, of fluctuation, of sudden

change, one sees here is strongly reminiscent of what one may see in other complex systems (chaos theory). It may be important, here, to consider migraine in this way, as a complex, dynamical disorder of neural behavior and regulation. The exquisite control of what we call health may paradoxically be based on chaos.[1]

In *The New Wounded*, Catherine Malabou redraws the relation between brain, mind, and soul, starting from the consideration that Parkinson's disease, Alzheimer's disease, and post-traumatic stress are now the prevailing forms of mental suffering, and that we should therefore be ready to link the neurological to the cognitive sphere, and the cognitive to the psychological sphere.

> Cerebral activity goes well beyond the mere work of cognition, and even of consciousness, to encompass the affective, sensory and erotic fabric without which neither cognition nor consciousness would exist.[2]

The theoretical foundation of Freud's psychoanalysis was essentially an act of separation of psychology from neurology, and this act resulted in the interpretation of psychic processes in terms of language. But "the unconscious is structured like a language," says Malabou, "only to the extent that the brain does not speak."[3]

The Freudian interpretation of sexuality is based on the suspension of any consideration of the physical brain, but Malabou questions this suspension, noting that:

> it does not take much—a few vascular ruptures, minimal in terms of their size and scope—to alter identity sometimes irreversibly. [...] Contrary to what Freud affirms, sexuality is always exposed to

a more radical regime of events: the shock and the contingency of the ruptures that sever neuronal connections. [...] From now on, people with brain lesions will form an integral part of the psychopathological landscape.[4]

Trauma, the physical decay of cerebral functionality, and stress provoked by external aggressions—these conditions cause transformations that have psychic implications. Malabou suggests a neuro-psychoanalytical approach in order to describe the psychological effects of neurological traumas and transformations.

According to Malabou, we need to integrate the neurological approach with the psychoanalytic approach if we want to understand organic diseases such as Parkinson's and Alzheimer's.

The same can be said of the cognitive mutation produced by technological transformation and the intensification of nervous stimuli proceeding from the accelerated infosphere. The hyperstimulation of attention, and the dramatic change affecting our mental environment can be considered a mutagenic factor belonging to the sphere of trauma, and therefore demanding an integration of psychological and neurological approaches.

The concept of neuroplasticity is crucial from two points of view: it provides the condition to understanding how the neural substratum is adapting to the cognitive mutation underway. But it also inaugurates the possibility of envisaging a conscious action to transform the social mind.

According to Dimitris Papadopoulos,

plasticity starts where the gene stops: the specificity of the individual organism. Plasticity appears when epigenetics is at work: the worldly making and remaking of the totality of an organism in the process of its development. Rather than the relative malleability of brain matter, plasticity now refers to the possibility of recombining brain-body matter. Not as an abstract and general process of neuronal regeneration but as a process that takes place epigenetically, that is according to the specific and contingent realities of each particular organism.[5]

The emergence of the body is not the mere deployment of the information contained in its DNA, but is the interaction between the genetic information and the environmental conditions in which the genes become the organism.

According to Papadopoulos, the emergent body exists in the realm of its own developmental trajectory and actuality, but it is emergent because the creation of new forms is always limited by the contingent conditions of existence. The emergent and embodied brain-body is unthinkable, and indeed cannot exist outside of the formative chronotope of ontogenesis. Papadopoulos' description of epigenesis, and his description of the emergent body are reminiscent of the concept of the body without organs that Deleuze and Guattari describe in *Mille Plateaus*. The neuroplastic description goes beyond the computationalist model.

Mental representations in cognitivism are the result of innate neurophysiological processes that are context independent and universal in the human brain. Thinking has universal algorithmic structure and resides in fixed neuronal architectures.

What is crucial in connectionism is that the weighting of the nodes is not given but emerges through learning. While computationalism presupposes innate neuronal structures, connectionism presupposes semi-open, non-linear architectures that unfold during the very process of ontogenetic development. Brain matter is simultaneously the actor and the result of its own activity.[6]

However, as Malabou argues in her book *What Should We Do With Our Brain?*, more important are the political implications of the neuroplastic model.

According to Malabou, plasticity is the relation that an individual entertains with what, on the one hand, attaches him originally to himself, to his proper form, and with what, on the other hand, allows him to launch himself into the void of all identity, to abandon all rigid and fixed determination.[7]

The adherence of mental activity to the neural substratum does not explain everything about the emergence of the conscious body, and the shift from neurology to consciousness can only be explained by referring to interaction with the environment and to the intentionality that is inscribed in consciousness.

In the words of neurologist Antonio Damasio, consciousness is "how the owner of the movie-in-the-brain emerges within the movie."[8]

Affection and Language

An imbalance between the organic potentiality of the brain and the effects of environmental neural stimulation affects the cognitive process as well as sensibility.

Memory, language, attention and the very ability of critical discrimination are involved in the process of mutation that is invading the organic features of nervous activity and interfering with established forms of cognition.

A mutation is invading the cognitive process at many levels, since mental activity is involved in the networked, digital infosphere.

Language is becoming more and more fragile, as a new generation of humans is learning more words from a machine than from its mothers.

In the book *L'ordine simbolico della madre*, the Italian feminist philosopher Luisa Muraro suggests that the access to language is made possible by the affective relation to the body of the mother.[9] The very relation between the signifier and the signifier is based on trust in the naming suggested by the mother. Words mean something because they have been performed and exchanged in the affective pragmatics of the bodily discovery of the environment.

One believes that this liquid is *water* because one's mother said so. The origin of the link between the signifier and the signified is not an operation, but occurs through affect. The voice, the pragmatic context, and bodily import are what establish the link of signification. The world is significant because it has been permeated by the affective creation of meaning.

Now, two different and converging processes of change happen at this level. The first is the separation of children from the body of their mother in the years of language acquisition. Women's emancipation has been turned into the capitalist subjugation of women's time and attention. Women are increasingly captured by the global labor market. Armies of women migrate from the poor cities of the global South to the busy metropolises of the affluent North. Millions of children in Manhattan, Milan and London are looked after by nannies who have come from Manila, Nairobi or Jakarta, and they spend more and more of their time interacting with screens. Language acquisition is therefore transferred from the affective environment of bodily contact, to the operational environment of the universal linguistic machine.

Words are affectively associated with meaning. As language acquisition is separated from the body, as it is reduced to the function of operational signification, the link between words and reality is weakened, and becomes frail, precarious. The meaning of words is reduced to an operational convention, devoid of bodily roots.

The singularity of language is rooted in the voice, the point of conjunction between meaning and flesh. Once words are separated from the voice, their effectiveness and meaning is then only based on the convention of operational effectiveness. The singularity of linguistic performance is lost, since language acquisition occurred in conditions of conventional conformity.

Sociologists and economists use the word precariousness in order to refer to the juridical transformation of labor relations. Workers become precarious when contracts and laws no longer protect them, when they must look for work continuously, and negotiate their own working conditions and salary. However, I think that the core transformation underlying the process of social precarization, and paving the way for the destruction of the links of solidarity between workers, is to be found in the psychological and cognitive sphere.

The weakening of language, its reduction to an operational mode, is the cognitive and emotional condition of the current process of precarization of life in social space.

The acceleration of the infosphere is affecting every facet of cognitive activity. In the networked environment, attention is continuously stimulated and mobilized, and the ability to focus on a singular flow of information is weakened, as demonstrated in the rise of Attention Deficit Disorders, a symptom of the weakening of the mind's ability to elaborate experience.

Consequently, the process of memorization is transformed as well. Humans tend to transfer their memory to machines, and to memorize essentially technical protocols to access that uniformed memory. The uniformity of memory, in turn, affects the imagination. For imagination is the recombination of memorized material.

The singularity of imaginative recombination is linked to the singularity of stored engrams, mnesic traces, or memes.

In the book *Meme Wars* published by *Adbusters*, Kalle Lasn discusses the automation of collective memory and the increasing uniformity of imagination.[10]

I want to make one last remark concerning the cognitive faculty of discrimination, the basis of the cultural attitude that we call critique. The ability to distinguish true and false in enunciations, emphasized by modern philosophy and political thought, was possible due to the prevalence of written text in the infosphere.

When exposed to sequential information, the singular mind can sequentially interpret the meaning of an enunciation, in order to discriminate about the reliability of its content.

As the electronic flow of information goes faster, time for elaboration shortens. The critical interpretation of the info-flow becomes increasingly difficult since the process of interpretation must go faster and faster.

For this reason, the critical faculty tends to be replaced by mythology, as McLuhan predicted in *Understanding Media*. Mythology becomes the prevailing mode of mental elaboration, since the simultaneity of configurational media replaces the sequentiality of the written text.

The Mind's We

In November 2012, I took part in a conference entitled *The Psychopathologies of Cognitive Capitalism*.[11]

The subject, the location, and the intellectual blend evidenced by the list of speakers suggest that this conference was a meaningful step in the creation of a philosophical and political awareness of the present crisis affecting the mental ecology of capitalist civilization, and its possible evolution. For the first time, to my knowledge, a European methodology particularly focused on the problems of social subjectivation confronted the Californian experience, and the specific composition of labor that is peculiar to the land of Disney, Apple, and Google.

In this conference, the theoretical field of social recomposition was approached from the conceptual point of view of neuroplasticity, which encompassed neurology, ecology of mind, and psychopathology.

Since the 1960s, in the Californian context, psychedelic experience and research on altered states of mind encountered the mind-changing potencies of the high tech industry. The Institute for Mental Research in Palo Alto, the works of Gregory Bateson and of Paul Watzlawick, the literary imagination of Philip K. Dick, and the psychedelic politics of Timothy Leary marked a mentalist reframing of the legacy of European philosophy and opened the way to a re-conceptualization of social becoming.

In the last thirty years, both the ideology of the new economy and the reality of new technologies found their cradle in California, nurturing a techno-mentalist culture, and giving rise to a special brand of social Darwinism that is expressed in cultural outlets, magazines such as *Wired* or books such as Kevin Kelly's *Out of Control*.

The technical transformation implied in the process of globalization is changing socio-cultural prospects so deeply that the conceptual tools inherited from European critical theory no longer suffice to imagine the future of human evolution. For this reason, I believe that the Los Angeles conference marked the first attempt to displace the object of socio-anthropological reflection, and to link the conceptual spheres of social recomposition and psycho-subjectivation with that of techno-mental evolution.

Sub-Individual and Super-Collective: The Mind's We

As Bruce Wexler recalled in his lecture during the conference, man is the animal who shapes the environment that shapes his brain. The infosphere is the environment where such mind-shaping occurs, and in the contemporary technosphere, politics, as the activity whose aim is to change institutions and collective behavior, is increasingly replaced by brain-altering devices. The problem of neuroplasticity comes to the fore in analyzing social change. The concept belongs first of all to the biogenetic domain, and refers in particular to epigenesis, the process that leads from the genetic information of DNA to the development of the organism. Contrary to the deterministic interpretation of biogenesis, researchers in neuroplasticity assume that the epigenetic process is neither inscribed nor pre-formed in DNA, and claim that an organism's epigenetic deployment is influenced by its environment. They therefore speak of neuroplasticity to assert the neural system's ability to adapt to the environmental conditions in which the organism evolves.[12]

My focus here is on the evolution of the social neural system, its ability to adapt to the external changes of the environment, particularly to the changes of the technical sphere.

In the present digital infosphere, conscious activity is involved in supra-individual connective concatenations. The connective concatenation shapes cognitive activity, and the unconscious according to a discrete—*versus* continuous—modality of perception. Syntactic rules of semiotic exchange replace those semantic rules that operated in the dimension of conjunctive relation and analogical communication. Those who dwell in the global digital sphere cannot escape the implications of connective concatenation if they want to interact in the collective sphere. If the individual wants to produce effects in the collective dimension, he or she must comply with its rules of interaction. This is why the art of politics has broken down, because one cannot interact efficiently in the collective dimension if one has not previously accepted the rules of compliance that shape language, action, and the interpretation of signs. The establishment of a connective format of interaction between humans reframes social composition. It is true that dissident thought is possible, and that dissident enunciation is also possible. But those uttering such dissident enunciations are in effect renouncing communication, because the format that makes communication possible is inept at conveying messages that are not compatible with the code.

In the industrial world, the social brain was modeled by standardized acts of physical production, but the mental sphere was only partially involved in the process of standardization. The metal worker of the classical industrial factory was forced to move his muscles according to the rhythm of the assembly line, but his mind was relatively free. Cognitive capitalism is all about the standardization of cognitive processes, and mental activity cannot be detached or diverted from the flow of information, since this flow is precisely the cognitive machine. Cognitive processes are directly shaped by connective formats, and unconscious activity itself is influenced by the overall transformation of our mental environment.

In *The Mind's I*, a book edited by Daniel Dennet and Douglas Hofstadter published in 1981, philosophers, psychologists, computer programmers and novelists were invited to investigate the interdependency between the hardware and the software of cognitive processes. The main question of the book could be expressed as follows: What is the relation between the neuro-physical composition of the brain and the conscious self-perception of the thinking organism? What is the mind of my *self?* What is the *I* of my *mind?*

In that book, Dennet and Hofstadter investigated the effects of processes of interaction on the individual brain. I believe we should now reformulate their basic question at a different level. We should investigate the effects of the connective infosphere on the social mind, and simultaneously on a sub-individual dimension, in order to understand how sub-individual flows are reshaping the collective space of sensibility. At the end of the day, we need to study the formation of the social mind from the point of view of the relation between sensibility and culture: the mind's *we.*

Such an investigation will question the limits and range of the plasticity of the neural system.

Structure and Machine

Operaisti Neo-Marxist Italian thinkers, and post-structuralist French philosophers such as Deleuze, Guattari, Lyotard and Foucault—notwithstanding the differences in their intellectual trajectories—investigated a similar question, that of the development of processes of subjectivation, starting from the material layers of social existence, technology, language, and affections. Thanks to the formulation of the concept of *composition* (class composition, social composition), Italian neo-Marxism in the 1960s directly questioned structuralist methodology. While structuralism is based on the notion that the subject's evolution is governed by internal patterns, according to a post-structuralist approach, social composition evolves due to the interference of external factors that enter the space of subjectivation. It is not internal structures, but external machines that shape molecular, sub-individual flows, funneling them into the provisional, ever-changing forms of sensibility.

In the article "Machine et structure," published in the journal *Change* in 1971, Félix Guattari explained his break with Lacan and his emancipation from the Freudian conception of psychoanalysis, which was based on the assumption that the unconscious could be described as a system of structures, linguistic, mythological, and symbolic.[13]

A structure posits that its elements exist within a system of references connecting every element to other elements, in such a way that the structure itself can be referred to other structures as an element in a larger structure. Such structural thinking through totalization and de-totalization grasps onto the subject, because it does not accept the idea of losing its grip on it if it cannot reinscribe it within a new structure.

The same doesn't apply to the machine, which we can describe as eccentric in relation to the subject. The machine comes from the outside; it pushes and displaces the subject, changing the landscape that surrounds it. Temporality enters the machine in a variety of ways and is the dimension in which events occur. The emergence of the machine marks a break that is not homogeneous with a structural representation.[14]

A structure is a system of inner relations, of interactions governing the subject from the inside. A machine, on the other hand, is an eccentric actor. It comes from the outside and changes the framework in which the subject is located, so that the subject itself changes form. The machine thus jeopardizes the structural pattern, and provokes a displacement of the subject. While structures are essentially morphostatic and territorializing, machines are factors of deterritorialization that lead to the generation of new forms.

The concept of the machine—that simply means whatever agent is working and producing effects—gives Guattari the possibility of replacing the structured subject with the vision of a process of subjectivation that belongs to the sphere of the event, not to the sphere of replication or repetition, because it is the effect of the action of machines.

Composition, Consciousness, and Subjectivation

There is a core investigation to the aforementioned French and Italian authors that centers around the following questions: What is the subjective side of social becoming? How do social forces develop different forms of consciousness and conflicting intentions? The answer can be found in the word *composition*. The subjective side of social becoming is in perpetual flux, since it results from the never-ending transformation of the psycho-cultural composition of the collective mind, the collective soul, and the collective unconscious.

The becoming of social subjectivity can be seen as a solution, in the chemical sense, a mixture of various substances melting together. Consciousness is the surface of the perpetual process of de-composition and recomposition that occurs within the social subconscious, and that interacts with the cultural limits of imagination and thought, as well as with the neurological limits of the brain. Consciousness is the ability to locate oneself within the map of the compositional flows of information, desire, conflict, etc. I propose to coin the term *compositionism* to encompass the two philosophical movements that flourished in Italy and France during the extended moment of 1968, and that have had long lasting-effects in contemporary philosophy. The philosophers from those movements retraced processes of subjectivation through the recognition of the perpetually changing composition of social life, via cultural, economic, psychological and mythological flows that all enter into the process of subjectivation.

In his last book, *Chaosmosis*, Guattari writes that, "among the fogs and miasmas which obscure our *fin de millénaire*, the question of subjectivity is now returning as a leitmotiv. [...] all the disciplines will have to combine their creativity to ward off the ordeals of barbarism, the mental implosion and chaosmic spasms looming on the horizon [...]."[15]

What does he mean by the expression *chaosmic spasms*?

The Chaosmic Spasm

According to a medical lexicon, a spasm is the sudden, involuntary contraction of a muscle, which generally involves pain. In the context of the analysis of social subjectivation, I would say that a spasm is an excessive, compulsive acceleration of the rhythm of the social organism, a forced vibration of the rhythm of social communication.

A spasm is a painful vibration that forces the organism into an extreme mobilization of its nerves and muscles. We should understand such acceleration, and its accompanying painful vibration, in the context of the contemporary sphere of cognitive work, and of its exploitation of the nervous system. I call this environment semiocapitalism, since the means of valorization inherent in it are essentially semiotic tools. Once cognitive energy becomes the main force of production, since capital valorization demands ever-increasing productivity, the nervous systems of the organisms existing under semiocapitalism are subjected to accelerated exploitation.

Guattari always conceived of technology as a factor of enrichment, as a means of enhancing the mind, and as an avenue for social liberation. But machines interweave and connect with capitalist exploitation, producing effects of subjection aimed at the continuous increase of productivity and exploitation.

This leads to spasms, effects of the violent penetration of capitalist exploitation into the sphere of information-technologies acting on cognition, sensibility, and the unconscious itself. Sensibility is invaded by the acceleration of information, and the vibration induced by the acceleration of nervous exploitation induces a spasm, or a spasmic effect.

What should we do when a spasm arises?

Guattari, it should be noted, does not use the word *spasm* in isolation. Rather, he invokes the term *chaosmic spasm*. For him, chaosmosis is the overcoming of the spasm, the relaxing of spasmodic vibration. In the interaction between the individual and the collective sphere, in the link between individual neural activity and connective concatenation, the *mind's we* evolves. The neuroplasticity of the individual organism interacts with the rhythms of the collective automatisms of the swarm.

Why does Guattari use the expression *chaosmic spasm*? How can the spasm be seen as chaosmic? And first of all, what is chaosmosis?

Chaosmosis is the creation of a new, more complex order (involving both syntony and sympathy) out of a situation of chaos that has emerged as the effect of the spasmodic acceleration of the semio-universe surrounding the organism. Chaos implies an excess of speed in the infosphere in relation to the brain's capacity of elaboration. Chaosmosis is the osmotic passage from a state of chaos to a new order. But the word *order* does not have a normative or an ontological meaning. Order, here, is the harmonic relation between the mind and its semiotic-environment, as well as a shared mindset: sym-pathy, a common perception.

In *What is Philosophy?*, a book about philosophy and also about growing old, Deleuze and Guattari speak of the relation between chaos and the brain. "From Chaos to the Brain" is the title of the conclusion of the book, and it begins with the following:

> We require just a little order to protect us from chaos. Nothing is more distressing than a thought that escapes itself, than ideas that fly off, that disappear hardly formed, already eroded by forgetfulness or precipitated into others that we no longer master. These are infinite *variabilities*, the appearing and disappearing of which coincide. They are infinite speeds that blend into the immobility of the colorless and silent nothingness they traverse, without nature or thought. This is the instant of which we do not know whether it is too long or too short for time. We receive sudden jolts that beat like arteries. We constantly lose our ideas.[16]

Since consciousness is too slow to process the information that comes from an accelerated world (info-technology multiplied by semio-capitalist exploitation), the world can no longer be translated into a cosmos, a mental order, syntony, and sympathy. What is at stake here has very little to do with politics and history, but much more to do with the neuroplasticity of the evolution of the brain.

Chaosmosis is a reframing of the relation between the infosphere and the mind, a process of re-syntonization and re-focalization that cannot be pre-arranged by political will. It can only be prepared by a modeling of sensibility.

Chaosmosis is the shift from a rhythm of conscious elaboration, to a new rhythm that is capable of elaborating what the previous rhythm could no longer consciously process.

In order to define the shift from one rhythm to another, from one refrain to another, Guattari proposed the concept of *chaoide*. A chaoide, in Guattari's parlance, is a sort of de-multiplier, an agent of re-syntonization, a linguistic agent that acts as a different refrain than the spasmic refrain, and that can de-multiply the spasmodic rhythm.

"The primary purpose of ecosophical cartography," writes Guattari, "is thus not to signify and communicate, but to produce assemblages of enunciation capable of capturing the points of singularity of a situation."[17]

Guattari's philosophical, political, and schizoanalytical project aimed at re-focalization and singularization, not adaptation. Indeed, the goal of prevailing psychotherapeutic techniques was to enable the suffering organism to adapt to its social and technical environment. The intention of psychopharmacology and psychiatric therapy was to soften psychic suffering in order to normalize behavior and to reduce the existing subjectivity to the tasks of cognitive exploitation. Reprogramming was a technique aimed at normalizing singularity, and at restoring neurotic subjection to the process of neural exploitation and the accumulation of capital.

Guattari's schizoanalysis is based on the idea that healing is a process of singularization, not one of conformity. But this process of singularization implies a complex dynamics of the mutual transformation of the social environment and of individual minds.

In the sphere of financial capitalism, the prevailing linguistic concatenations produce a spasmogenic rhythm. Not only does the spasm exploit men and women's work, subjecting cognitive labor to the abstract acceleration of the info-machine, but it also destroys the organism's sensibility by subjecting this sensibility to the stress of competition and acceleration. In Guattari's parlance, a chaoide is a semiotic device that allows the organism to disconnect from a pathogenic rhythm, and enables the creation of a new concatenation between consciousness and the infosphere. Chaosmosis is the evolutionary process of recomposition that leads to the emergence of a new concatenation, and therefore to the possibility of a new sympathetic syntony of the molecules composing the social body with the flows circulating in the infosphere.

The first step in this chaosmosis will be to disentangle from the present, stressful concatenation. The second will be the neural reframing of the relation with the infosphere. This is not a political project, since politics has broken down and is unable to deal with the processes of meta-subjectivation that are implied in chaosmosis. Guattari suggests that we must create chaoides for disentanglement, for prefiguration and re-syntonization. Chaoides have nothing to do with the sphere of will and political decision. Instead, they belong to the sphere of art, education, and therapy, where sensibility is shaped.

The Paradigm Shift and the Cultural Transition

During the transition from industrial capitalism to semiocapitalism, the transformation of production into immaterial production invaded the process of subjectivation. Precarization and the fractalization of labor provoked a deep mutation of the psychosphere that resonated with technological and cultural becomings. This mutation has not followed a linear process, since the different levels of human activity, be they cultural, psychological, and neural, do not change in unison, but occur according to different temporalities and rhythms of transformation.

In recent decades, as the transition to semiocapitalism was underway, many authors, from many different points of view, spoke of a paradigm shift at the epistemological, technological, and economic levels of social life.

A short reminder of the most important theorizations of this notion of a paradigm shift may be useful.

In his most celebrated book, *The Structure of Scientific Revolutions*, the epistemologist Thomas Kuhn defined the transition from one epistemological framework to another in terms of a paradigm shift. This concept is essential in order to analyze the transformation that has invaded every sphere of theory and of social activity in the age of the transition from the sequential to the simultaneous technosphere.

In the book *Understanding Media*, published in 1964, Marshall McLuhan had already spoken of electric light as a medium that redefined the notion of content.

> As electrically contracted, the globe is no more than a village. Electric speed in bringing all social and political functions together in a sudden implosion has heightened human awareness of responsibility to an intense degree. [...] The instance of the electric light may prove illuminating in this connection. The electric light is pure information. It is a medium without a message [...]. For the "message" of any medium or technology is the change of scale or pace or pattern that it introduces into human affairs.[18]

First electricity, then electronics: this is the technological background of the transition that occurred in late modernity and gave way to a general transformation of the economy, social communication, and culture.

L'informatisation de la societé (*The Digitalization of Society*), a report commissioned by French president Valéry Giscard D'Estaing, was published in Paris in 1977. In this book, sociologist Alain Minc and engineer Simon Nora anticipated telematics—a major technological innovation proceeding from the intersection of the telephone with computing—and predicted the upcoming crisis and the deterioration of national sovereignty following the globalization of political and economic information.

Alvin Toffler, in *The Third Wave*, first published in 1980, forecast the overall transformation of the social sphere through the spread of electronic technologies into daily life and the economy.

> The Third Wave brings with it a genuinely new way of life based on diversified, renewable energy sources; on methods of production that make most factory assembly lines obsolete; on new, non-nuclear families; on a novel institution that may be called the electronic cottage; and on radically changed schools and corporations of the future. The emergent civilization writes a new code of behavior for us and carries us beyond standardization, synchronization, and centralization, beyond the concentration of energy, money, and power.[19]

But the paradigm shift also concerned scientific and epistemological space. According to Jean-François Lyotard, "the status of knowledge is altered as societies enter what is known as the post-industrial age and cultures enter what is known as the postmodern age."[20] In the field of science and epistemology, in fact, the mechanical paradigm that had prevailed in the centuries following the Newtonian revolution is now giving way to the paradigm of uncertainty outlined by Werner Heisenberg in 1959.

However, the paradigm shift must be conceived as a tendency and a possibility, not as a necessary consequence determined by a change in the social and technological environment. In fact, the tendency and the possibility implied by the transformation of the environment are obstructed and hindered by the persistent force of the past. This is why the process of reconfiguring human subjectivity and the neuroplastic social brain is asymmetrical and asynchronous. Capitalism, as the general form of the economy, entangles and obstructs the potential developments of the social brain. If we want to understand this entanglement, we must retrace the history of contemporary subjectivity, and analyze the various layers of its becoming, psychic, cultural and aesthetic.

The concept of neuroplasticity is ambiguous. It can signify a mere re-adaptation of neural activity to the needs of the technosphere and of semiocapitalism, implying biopolitical formatting and the submission of the mind to the economic goals of cognitive labor captured and subsumed by semiocapital. But we can also think of neuroplasticity as the chaosmotic reframing of neural activity, enabling the emergence of a post-capitalist paradigm that cannot be expressed in conscious and rational terms, but emerges through chaos. From this point of view, neuroplasticity means a re-tuning (a new syntonization) of the neuroplastic brain with an environment that the contemporary brain only perceives and detects as chaos.

Modern culture has conceived of change in terms of history, and of action in terms of politics. I think we should focus on neuro-evolution, and develop forms of action aimed at consciously shaping it. The main task before us now is to find ways to consciously interact with neural evolution, and de-activate those techno-linguistic automatisms that entangle the mind's activity with the established frame of connective capitalism. We have to think in terms of meta-connectivity.

Social Morphogenesis after the End of Democracy

Morphogenesis is the process of becoming form. It can occur due to the conscious and voluntary action of a subjective morpho-generator. Or, it can be the self-organization of chaotic matter that cannot be governed by will or by consciousness, the spontaneous emergence of a form that finds a way towards a new order.

Do new social forms come into existence spontaneously, as the effect of natural processes that humans must accept and deal with? Or do they come into existence because of a conscious act of will, as the effect of a project, a conflict, or a political decision and action?

This dilemma is obviously too simplistic. Things happen in different ways; spontaneity and will are sometimes impossible to distinguish; chance and intentions intermingle in the historical process. However, I want to know, in our contemporary conditions of crisis, if the process of change can be managed by deliberate human action. My question is: can the contemporary agony of modern capitalism, that is, the other, paradoxical face of the contemporary triumph of capitalism, be consciously managed by political action, and turned into a new form of social existence?

Obviously, political power has never actually been able to control the entirety of social relationships, and reason has never been able to reduce the infinite complexity of reality to knowledge. Yet, in modernity, it was possible to know and to control a relevant part of social complexity. In retrospect, modernity can be seen as an age in which the social world was relatively reduced and governable: the technique of politics was based on the ability to know what was relevant in the overall flow of information, and politics was the art of predicting the becoming of information so as to govern the main events and trends.

The shift to the hyper-complex reality of the networked world has made it impossible to understand and to control the relevant flows of information that circulate in the infosphere and that continuously stimulate the social brain. The old art of politics is thus increasingly impotent to predict, to govern, and to direct collective action towards a common goal. As a consequence, power is less and less reliant on the possibility of government, and instead, attempts to subject bodies and the flow of information to the model of governance.

Governance is the effect of the embedding of automated techno-linguistic devices in the continuum of social behavior and communication. Thanks to the automation of interconnected relations, organization and order are expected to emerge from the chaotic matter of social life, that is, from biology, media, technology, economic interests, affections, and the unconscious. Life, language, and production are penetrated and interconnected by a dissemination of devices (*dispositifs*) that aim to subject linguistic behavior to pre-conceived procedures and finalities, devices whose task is to make actions and enunciations predictable and manageable, such that, in the end, they are reducible to the overarching goal of capitalist accumulation and expansion.

In 1972, German philosopher Jürgen Habermas and sociologist Niklas Luhmann debated the effects of the evolution of media and the expansion of social information.

Although I am over-simplifying, I will roughly summarize their complex arguments here. While the optimist, rationalist Habermas believed that the enhancement of human communication was destined to consolidate democracy and to improve daily life, Luhmann expressed doubts about the positive effects of the expansion of the infosphere, and conjectured that this expansion could lead to a shift in emphasis from rational, political decision-making to administrative processes of differentiation. The expansion of the infosphere would not reinforce democracy, but would endanger it.

In our contemporary situation, marked by the infinite expansion of the infosphere—and in particular of economic and financial information—rational elaboration and political decision-making are no longer within the reach of individual or social organizations. As a result, democracy has been replaced by automatic procedures, by algorithms and devices for automatic selection and recombination whose general rationale is the replication of the capitalist form.

The efficiency of rational will and political action depended on the possibility of processing information flows and choosing the best options. When information flows become too fast, too dense, and too compressed for rational elaboration and for on-time decisions, then what was called politics in the modern age becomes unable to choose and to consciously generate new forms.

Neither democracy nor authoritarian power seems able to process the infinite and hyper-accelerated flow of information. Morphogenesis can therefore no longer be a process of conscious decision-making and elaboration, and turns into a self-regulating effect of blind emergence. The info-networked super-organism is now evolving outside of the sphere of human decision-making and knowledge, this, although its evolution affects the human environment—sometimes in a catastrophic way.

Capitalism as Double Bind

Slavoj Zizek has said that financial collapse would not be the end of the world, but that it would simply be the end of capitalism, something that we can hardly imagine. He may be right, but the problem is this: what if the semiotic model of capitalism—and particularly of financial capitalism—has become our only grid of perception and interaction? What if the biopolitical model of capitalism has pervaded the very fabric of social reproduction?

"The organism which destroys its environment destroys itself," stated Gregory Bateson in a speech delivered at the Korzybski Memorial in 1970. And he continued:

> If I am right, the whole of our thinking about what we are and what other people are has got to be restructured. This is not funny, and I do not know how long we have to do it in. If we continue to operate on the premises that were fashionable in the pre-cybernetic era, and which were especially underlined and strengthened during the Industrial Revolution, which seemed to validate the Darwinian unit of survival, we may have twenty or thirty years before the logical *reduction ad absurdum* of our old positions destroys us.[21]

Thirty years have gone by, and we have not stopped thinking according to the Darwinian terms of competition and of the survival of the fittest. Capitalism has triumphed, society is agonizing, and the planet is being dragged into agony as well. Monetarism as the autonomization of monetary function can be considered the capitalist attempt to escape or at least defer the final collapse. In order to prolong its survival, capitalism is subjecting social energy and the general intellect, in particular, to the obsession of financial stability and to the infinite growth of accumulation. The relation between capital and the general intellect (the living force of knowledge and the technological potency of labor) can be interpreted in terms of the relation between form and content.

Try to imagine this scenario, which is not so unlikely: Europe's financial system totally crashes, nation states stop paying wages to public workers, and, all of a sudden, money loses its grip on the social mind. Would our skills, our knowledge, and our competences be cancelled by this sudden, apocalyptic event? Not at all, of course. We would be the same as we are now. Engineers would be able to build bridges, doctors would be able to heal sick people, and poets would be able to create their imaginary worlds. Exactly as it is now, and possibly better.

The crumbling of the form would not affect the content.

But the agony of capitalist form, if it is protracted in time, will slowly but steadily dismantle social content, and it is already doing so. The de-financing and privatization of public schools, for instance, is going in this direction. It marginalizes the majority of the population, and leads to massive de-schooling. Private schools, on their end, reduce knowledge to the idolatry of economic dogma, technical skills are separated from social understanding, and science is separated from the humanities. This is the educational reform that started in 1999 with the signing of the Bologna Accords in Europe—which is in fact the destruction of the legacy of the autonomy of knowledge and of trans-disciplinary research.

Form is destroying content. The relation between form and content can at times become pathogenic, and produce what Gregory Bateson labeled a *double bind*.

Bateson described this process in his research on the genesis of schizophrenia. The context affects the understanding of a message in such a way that communication itself becomes jeopardized, since the form of interaction (the medium) changes the content, or perverts the understanding of the content (its meaning).

The more serious and conspicuous degree of symptomatology is what is conventionally called schizophrenia. [...] The literal is confused with the metaphoric. Internal messages are confused with external. The trivial is confused with the vital. The originator of the message is confused with the recipient and the perceiver with the thing perceived.[22]

In situations like these, content has to be disentangled from form, and a new form has to arise from the self-organization of content.

If we do not manage to disentangle the potency of the new technology from the old paradigmatic framework, the effect will be catastrophic because this potency will explode against society.

Entanglement is a concept that has nothing to do with Hegelian contradiction. It's not a problem of opposing forces, of struggle, overpowering, and overcoming (*aufhebung*, as Hegel says). Rather, it is a problem of the possibility contained within a form, and hindered or perverted by this form. Not only does entanglement impair the deployment of content, but it also transforms, and perverts the content of knowledge.

I use the word *entanglement* in order to refer to the capture of the social process of production, knowledge, and communication by the semiotic code of growth and accumulation. Marx calls this *real subsumption*.

Disentanglement would be the emancipation of content from the form that contains it, and would imply the full deployment of the potencies belonging to social knowledge.

The problem of disentangling content from form can be described as a problem of schizoanalysis, in the words of Félix Guattari. Schizoanalysis entails the ability to disentangle mental contents and psychic activity from the obsessive refrains that entangle the activity of the mind.

Generation by Schism

What is form (*morphè*)?

For Gestalt theorists, such as Wertheimer, or Koffka, the brain perceives objects thanks to the existence of forms that are embedded in the perceptual constitution of the mind. As *Gestalt*, the form is the condition for the differentiation of objects from the surrounding environment, and for the interpretation of their meaning. But we can also say that form is the semiotic pattern that our mind projects onto the world and the model for the generation of external objects. In this sense, form can also be defined as a prototype, the original in-formation that shapes matter. From this point of view, we can say that form is a general *semiotizer*. It is the paradigm that makes the attribution of consistent meaning to phenomena, which we experience as signs, possible.

There is a remarkable relation between the concept of *form* as semiotic generator, and Guattari's concept of *refrain*. I would say that the refrain is the sensibilization of a form, the translation of a formal prototype into a sensible subjectivity.

When Guattari speaks of chaosmic spasm, he is saying that the general semiotizer is no longer able to semiotize, and that the social mind is deprived of the ability to process information in a consistent way. The refrain is no longer able to resonate with the surrounding informational environment, since the semiotic generator (form) is mutating and shifting. Consequently, a sensible subjectivity is unable to grasp meaning, to interact semiotically, to concatenate, and suffers from this dystonia. This is the spasm. The organism has lost its syntonic relation with its environment; organic vibration no longer interweaves harmonically with the surrounding vibration, so the vibration becomes frantic and painful.

A process of re-syntonization then begins, which Guattari names *chaosmosis*. In the chaosmotic process, forms emerge from chaos through a vibratory approximation to an order that is functional and aesthetic. Functional order makes the manipulation of objects possible, while aesthetic order enables the conjunction of objects in the sensibility and sensitivity of the living organism.

In certain cases, the form, as semiotizer, can become a tangle, a generator of double binds. In these cases, the form (the *gestalt*) does not help interpret and deploy the contents of the collective mind, but rather limits the possibility of the development of matter-content, and jeopardizes the concatenation of the mind and the infosphere.

When form destroys content—knowledge, skills, technology, and social emotionality—the only alternative to devastation and death is to disentangle content from form.

In our society, capital is the general gestalt of our experience of the production and circulation of things, and is the general semiotizer. It is the source of meaning and the operational measure of those things that humans need, produce, and exchange. Capital, as form, has become a tangle for the development of the potencies implied in the general intellect. Only the removal of this form, the replacement of accumulation and growth by different paradigms of production and exchange can allow the general intellect to fully deploy its potencies.

From this point of view, the neo-Marxist emphasis on the potency of the general intellect, and the incompatibility of its full deployment with the persisting prevalence of capitalist form, matches the theory of the paradigm shift. But the authors who discuss the paradigm shift have generally forgotten to say that this shift is neither necessary nor linear. It is a tendency and a possibility, but in the process of social becoming, this tendency and possibility may be obstructed and hindered by the overwhelming force of the automatisms implied in past forms of subjectivity. We must be aware of this contradiction if we want to understand the real process of social morphogenesis in the present, and under present conditions.

Bateson uses the word *schismogenesis* to describe the process of differentiation internal to human groups, and the generation of new levels of anthropological integration. I want to use the same word—schismogenesis—in order to describe the separation that leads to the generation of new forms.

The generation of a form that is more likely to foster the development of the potencies of content can only happen through the dissociation and disentanglement of the potentiality of content from the form in which it is entangled. I therefore take schismogenesis to mean the self-organization of contents after their dissociation from their entangling form, and the proliferation by contagion (affective, informational, aesthetic contagion) of the new form generated by the schism.

Under our present conditions, several questions arise. Is disentanglement still possible when the mind of the social organism has been deeply infected by the viral proliferation of double binds? And what is the origin of the proliferation of double binds in the social mind?

The paradigm shift is a general tendency inscribed in the evolution of the contents of knowledge, technology and social production. But entangling forms hinder this tendency, acting as repetitive semiotizers and generating double binds in the social mind.

In order to understand why the social mind is unable to free itself from these entangling forms of semiotization, we must analyze the cultural and psychological becoming of sensibility, as well as the infection of the social organism that is invading the sphere of sensibility.

8

The Transhuman

Humanism, Technique and Language

In *The Postmodern Condition*, Jean-François Lyotard announced the decline of what he called *grand narratives*, the ideological tales that played a crucial role in motivating historical action throughout modernity. But I don't believe that the Lyotardian narrative about the end of narratives is true. New narratives—no less grand than the old ones—have taken hold of the contemporary imagination. Artificial intelligence is certainly one of them. But in the same book, Lyotard wrote things that in my opinion were more interesting. For instance, he said that a new status and structure of knowledge appeared when the technological system for production and transmission of knowledge was changed. This new status of human knowledge questioned the distinction between human and intelligent machines.

In a collection of essays published in 1991 entitled *The Inhuman*, Lyotard reflected on the blurring of the lines between humans and machines. In order to assess that the gap between human thought and artificial intelligence was unbridgeable, he resorted to the metaphor of the cloud: while artificial intelligence was based on rigorous consequentiality, the nature of thought, in Lyotard's terms, possessed the same indeterminability as the process of the formation of clouds.

Computers do not know what an event is.

What is a place, a moment, not anchored in the immediate *passion* of what happens? Is a computer in any way here and now? Can anything *happen* with it? Can anything happen *to* it?[1]

By using the word *inhuman*, Lyotard put on the table a much-debated question: is humanism in danger in the age of intelligent machines?

The saying attributed to the Greek philosopher Parmenides, "man is the measure of all things," is reminiscent of the words of Confucius, who believed that "it is man that makes truth great, not truth that makes man great."

The meaning of Parmenides' phrase, as well as that of Confucius, can be better appreciated if we refer to what Heidegger writes in the first page of his *Letter on Humanism*:

> Language is the house of being. In this home human beings dwell. Those who think and those who create with words are the guardians of this home. Their guardianship accomplishes the manifestation of being insofar as they bring this manifestation to language and preserve it in language through their saying.

And further in the same text:

> Ek-sistence can be said only of the essence of the human being, that is, only of the human way "to be."

As far as our experience shows, only the human being is admitted to the destiny of ek-sistence.

As ek-sisting, the human being sustains Da-sein in that he takes the *Da*, the clearing of being, into "care." But Da-sein itself occurs essentially as "thrown." It unfolds essentially in the throw of being as a destinal sending.[2]

The relation between language and technique is a good viewpoint for questioning humanism today, according to Heidegger. In his opinion, in fact, technique has become the true subject of language, the concatenation that utters enunciations. The technological dependence of language, and the incorporation of technical machines into the very generation of language are changing the nature of human-ness as never before.

The Heideggerian word *Da-sein* can be translated as *being there*, and means being in a situation of singularity, being as an event. Existence is an incidental (unnecessary and non-compatible) condition of eventuality, and the language of existence is the language of conjunction, the un-repeatable meeting between singularities.

The submission of language to technical rules of compatibility has transformed linguistic performance into connection. Humanism is at stake in this transition from the conjunctive to the connective mode of language.

Humanism as Indetermination

The Old Testament conceived of an impassive point of view of nature, and an emotionless flow of a-historical time. The impassive God who created man without feeling his suffering played a game that had nothing to do with men and women's earthly game. God and Satan felt no empathy whatsoever for the pawns of their chess game, as poor Job learned at his own expense.

The essential innovation of the New Testament was that God became Man, and came to earth to feel and suffer the same passions and pains that human beings were accustomed to feeling.

Thus the passionless time of God was interrupted, broken, and interwoven with the time of man. This is why the modern humanist revolution occurred within the space of Christianity, because both Christianity and humanism conceived of history in a temporal sphere that was not the sphere of eternal truth, or of impassible nature. Rather than the point of view of an impassive nature, man's suffering was the foundation of truth, as Confucius also believed.

It is not truth that makes man great, but man who makes truth great.

The foundations of modern civilization are to be found here, in the untwisting of a separate sphere—the sphere of human history and of human sociality—that does not comply with the eternal rules of the universe. Thanks to the scientific revolution, we know the mechanical laws that govern the planets, the sky, and stones, and we can disentangle a different law for human existence, one based on love and compassion. Indeed, the etymology of the word *compassion* is *cum-pati*, to suffer together with.

In the humanist sphere of modernity, after the separation from natural time ruled by the unchangeable laws of physics and from historical time ruled by the will of the prince or by democratic will, the establishment of political law was based on the understanding of human interests and passions. Modern civilization was based on the idea that the social world must not comply with the laws of the universe, but with the laws of compassion: mutual understanding, solidarity.

In *Oratio de dignitate homini*, Pico della Mirandola was explicit on this point. God had created man in a way that was different from the rest of the universe. The universe was built according to precise rules, but man had no built-in rules.

> We have given you, O Adam, no visage proper to yourself, nor endowment properly your own, in order that whatever place, whatever form, whatever gifts you may, with premeditation, select, these same you may have and possess through your own judgment and decision. The nature of all other creatures is defined and restricted within laws which We have laid down; you, by contrast, impeded by no such restrictions, may, by your own free will, to whose custody We have assigned you, trace for yourself the lineaments of your own nature. I have placed you at the very center of the world, so that from that vantage point you may with greater ease glance round about you on all that the world contains. We have made you a creature neither of heaven nor of earth, neither mortal nor immortal, in order that you may, as the free and proud shaper of your own being, fashion yourself in the form you may prefer. It will be in your power to descend to the lower, brutish forms of life; you will be able, through your own decision, to rise again to the superior orders whose life is divine.[3]

Modern history took place in this space of indetermination, and therefore of freedom. This freedom was not lawless, but human laws were a human construction, not the reflection of natural rules imposed by God.

Pico's vision reflected the humanist approach to the problem of freedom, and in it lay the thread that linked humanism and the Enlightenment. In the humanist space of indetermination, human reason created its own rules, and the universality of moral and political laws was based on human reason, not on natural law.

The Dismantlement of Social Civilization

Socialism, as the culmination of the modern construction of a humanist civilization, can also be considered a logical development of the humanist trajectory. Humanism affirmed the autonomy of human space from the impassible laws of nature, and the Enlightenment acted as a rational regulator of this autonomous human space. Socialist thought of the nineteenth century affirmed the possibility of justice and equality that would not be based on nature, but on human reason and compassion: the ability to share the same feelings, the same suffering, and the same goals.

As a result of these progressive developments, modernity culminated in the creation of a form of social civilization, a civilization in which common needs prevailed against the affirmation of individual interests. Social civilization aimed at preventing (although not always successfully) a war of all against all thanks to the establishment of a non-natural law.

Today however, the construction that I name social civilization is crumbling under the attacks of techno-financial capitalism, advancing under cover of social Darwinism. And the crisis of social civilization is leading to the crisis of humanism itself, since it is cancelling the humanist distinction between the kingdom of nature and the republic of men, and consequently the distinction between history and evolution.

Social-Darwinists say that benevolent principles cannot stop the affirmative strength of evolution, both in the social and natural spheres. If natural evolution is marked by the survival of the fittest in the dangerous environment of the planet, historical evolution is no exception. It is therefore useless to resist the prevailing force of the fittest for the sake of the socially weak.

Implicitly reclaiming the cancellation of any distinction between social life and nature, the philosopher and economist Friedrich Hayek asserts that Adam Smith's invisible hand regulates the market as a natural force.

According to the idea that only the fittest survive, and that the unfit are doomed to fail, neoliberal ideology obliterates the humanist distinction between the sphere of natural law and the sphere of moral reason. Since human relations—namely economic relations—follow natural laws of self-regulation, there is no need for special, regulatory interventions on the part of nation states or other political entities.

No special privilege is assigned to human kind. The animal spirits of economic agents are the only effective regulators of economic life, and therefore of social life as a whole.

The unleashing of the animal spirits of capitalism and the resulting dismantlement of the institutions of social civilization—two processes that have been underway over the last thirty years on a global scale—prove that reason and law are frail protections against the unbridled violence of capital.

It is time to acknowledge that social morality and the universal ethical law of the Enlightenment will never be efficaciously reinstated against the overwhelming force of corporate interests, since the process of globalization has in fact swept away almost all the regulations intended to protect human dignity and the natural environment.

The worker's movement and democratic forces were overly confident in the effectiveness of reason and lawfulness. From a materialistic point of view, we must understand that only the organized force of society can resist the organized corporate force of individual greed. The problem is that it is not easy to discover what force means in the social field.

The worker's movement tried to fight capitalism with armed revolutions, but this did not work well in the long run, as we know. The socialist state and the Red Army took the place of the oppressors, and in the end, most people preferred corporate exploitation to the dictatorship of the communist state.

The crisis of humanism began once the indetermination that Pico della Mirandola evoked in his *Oratio* came to an end through the establishment of a deterministic rule within the generation of language itself. If human liberation from natural domination started with the establishment of technique, and with the opening of a historical space of self-rule, technical development itself is creating the conditions for a return of determinism in language and social exchange. God gave humans the freedom to define themselves. Language was the space of this self-definition, and technology the instrument for making that freedom effective.

But we are now witnessing a paradoxical reversal, since technology is taking the place that the humanist God had decided to leave empty. Technology is replacing the deterministic God that God had decided not to be, because it is transforming language into a chain of automatisms, and therefore cancelling the indetermination that was the condition of possibility of self-definition and freedom.

Heidegger wrote that language was the house of being, but he also wrote that language belonged more and more to technique. When language is reshaped by connective technology, then the linguistic creation of being becomes regulated by mathematical, algorithmic chains. Event and being diverge, and singularity is cancelled.

In this de-singularization and dis-eventualization of being, Heidegger sees the advent of nihilism, as the final essence of modernity:

[...] humanity sets in motion, with respect to everything, the unlimited process of calculation, planning, and breeding. Science as research is the indispensable form taken by this self-establishment in the world; it is one of the pathways along which, with a speed unrecognized by those who are involved, modernity races towards the fulfillment of its essence.[4]

Humans began the process of the mathematization of language, and this process will cancel the possibility of freedom from the automatic chain of computational language. The space of being, that God decided to deliver to men as an empty space, is now filled by the generative power of the technosphere such that digital conventions become the *nature* of language, and that the digital nature of language halts humanist history, that is, history itself.

Evolution is back. And it is replacing the now *no longer humanist* human becoming.

History and Evolution

How can we distinguish the concept of evolution from the concept of history? Let's say that it is about scale, the scale of space and time. We don't speak of history when we think of the solar system, or intergalactic space, because the scale of space is out of human reach. We don't speak of history when we refer to atoms and sub-atomic particles because they are beyond the reach of political action. It is obvious that extraterrestrial events can interfere with history. If a meteorite destroyed the city of London, it would certainly have historical effects, but it would not be a historical event in itself.

The same can be said about the temporal scale of events. We are accustomed to thinking in terms of historical time when the rhythm of events can be scrutinized by the rational mind, and can therefore be influenced by political will.

Long-term transformations of the earth's surface are out of historical reach, although we can act historically and politically on carbon emissions, and, in the long run, influence global temperature. Micro-temporal events are beyond the sphere of human knowledge and control.

Only those events and bodies that are neither too large or too small, nor too fast or too slow to escape the human grasp can be objects of historical action and political will. What is too large or too small, too fast or too slow to be visible, perceivable, and manageable belongs to the sphere of evolution, not of history.

Scientific thought and technological change gave humans the ability to deal with those spatial and temporal dimensions that cannot be scrutinized by the naked eye, and that cannot be checked and subjected to rational discussion and critical decision. For this reason, we are exiting the dimension of history, and our actions must increasingly face an evolutionary appreciation.

Take the relation between political decision-making and financial capitalism. Thanks to the electronic transfer of digital data in global networks, markets have been hijacked by trading robots, many of which are so self-directed that humans can't predict what they'll do next. High-speed traders have overtaken the stock market because they can trade stocks in and out thousands of times per second. According to Scott Patterson, the new financial players moving into artificial intelligence are on the verge of tipping the entire system toward a global meltdown that could occur in minutes, maybe even seconds. Fiber-optic cables circle the earth, linking all financial markets at ever-higher speeds in the global algorithm race. Making political decisions takes days, sometimes months, and implementing political decisions in the dense fabric of social life takes years. Social change occurs in a dilated time in which actions are necessarily slow, since they must take into account cultural resistance and so on. Resistance through political movements is based on the slow time of discussion, persuasion, and social organization. Financial transfers that can jeopardize social life and change the political landscape overnight occur within the time span of nanoseconds. This is why the synergy of financial capitalism and digital technology has broken down political defenses against transhuman factors of change, and has propelled social life into a temporal dimension that is out of control.

In the meantime, finance, no longer consigned to the margins of economic life, is at the core of the creation (and destruction) of wealth. According to Christian Marazzi:

> You know that there is a whole tradition, both in orthodox Marxism and Keynesianism, or anyway in something linked to neo-Ricardian visions of finance, that considers finance as a deviation from real productive capital. I don't think that we can talk about finance in this way anymore. This doesn't mean that finance has become sexier. Now it is as painful and awful as it's always been.
>
> This is not the point: the point is that there is a transformation of the categories used in the 20th century. For instance it is difficult today to make a distinction between profit and rent, there is a becoming rent of profit, the dividing line between them is falling apart, but this is not because capital is not accumulating or growing anymore and so everything is just speculation, but because this is a new mode of production in which the relation between capital and labor has changed.[5]

By inserting nano-temporal events into the very texture of the process of social reproduction, financial capitalism is transforming our perception of time, and is forcing us to abandon historical temporality in favor of an evolutionary way of perceiving and expecting—and simultaneously is forcing us to abandon the political attitude and the expectation that human will can change something. The perception of time is shifting from the historical to the evolutionary mode. Someone could obviously object that technical devices and financial procedures are the product of human will and of social interests, which is true. Nevertheless, once the effects of voluntary action condense into automatisms, they take the form of a necessary concatenation that conscious will can no longer change, contest, or undo. The process of capitalist abstraction has progressively eroded the potency of concrete activity: digital financialization constitutes the final limit of this disempowerment and the economic framework of a biopolitical transformation that forces cognitive activity to mutate and that shapes the physical matter of the neural substratum itself. This transition from the sphere of historical humanism to that of evolutionary automatism can be described as building a kind of neuro-totalitarianism. The cognitive mutation induced by digital technology is a path in that direction.

The Transhuman Utopia

At the end of the nineteenth century, the Russian evolutionist Nikolai Fedorov anticipated many of the themes that would be developed in the first decades of the twentieth century, particularly in Russia and Italy, the two most backward countries of the continent, where, not surprisingly, the cult of the future excited the artistic imagination. Fedorov's utopia was based on the assumption of a boundless perfectibility of the human race, and aimed at opening the doors of the future to the colonization of space, the revival of the dead, and immortality.

Persuaded that the evolutionary process was directed towards increasing intelligence, and that human intelligence could progress without limitations, Fedorov set the goal of abolishing death as the common cause of a future mankind.

Human beings, in Fedorov's view, were dying because they had been unable, so far, to regulate their organisms' psychophysiological processes. The rationalization of nature and the medical regulation of viruses and epidemics would give humans the ability to become free of the *error* of death.

The energy of the cosmos—in Fedorov's utopia—was to become the inexhaustible source of human life and historical progress.

Fedorov's cosmist philosophy was only the first of many transhumanist utopias that punctuated the twentieth century, and that often turned into dystopian nightmares.

Scientists often expressed their concern about the ambiguous features of techno-utopianism. Norbert Wiener, who coined the word cybernetics and who can be considered the father of computational technology, was conscious of the dangers intrinsic in the thinking machine, since he knew that computers might step beyond the reaches of human control, or become the tools of greedy capitalists. He therefore tried to create the conditions for independent institutions for scientists.

In the field of technological innovation, utopia and dystopia grow together, particularly when the faculty of cognition is challenged by the possible developments of the computing machine.

When, at the turn of the century, the synergy of biotechnology and artificial intelligence opened the way to unprecedented opportunities to replace the organic body with a synthetic organism—once the miniaturization of digital devices paved the way to nanotechnology and thus to the insertion of technical automatisms into the very fabric of life—utopia and dystopia came closer, almost melting into the transhuman imagination of the future. Two names exemplify the two diverging directions and opposite possibilities that are implicit in the transhuman technological imagination: the name of Ray Kurzweil and the name of Bill Joy.

Computer scientist and co-founder of Sun Microsystems, Bill Joy is the author of an important article that was published by the magazine *Wired* in the year 2000. The article, entitled "Why the Future Doesn't Need Us," raised critical questions about the unintended consequences of genetic engineering and nanotechnology, whose imminent deployment threatened to pervade the natural environment, and the human brain itself.

Bill Joy began his article by discussing his ethical concerns:

> From the moment I became involved in the creation of new technologies, their ethical dimensions have concerned me, but it was only in the autumn of 1998 that I became anxiously aware of how great are the dangers facing us in the 21st century. I can date the onset of my unease to the day I met Ray Kurzweil, the deservedly famous inventor of the first reading machine for the blind and many other amazing things.
>
> Ray and I were both speakers at George Gilder's Telecosm conference, and I encountered him by chance in the bar of the hotel after both our sessions were over. I was sitting with John Searle, a Berkeley philosopher who studies consciousness. While we were talking, Ray approached and a conversation began, the subject of which haunts me to this day.

I had missed Ray's talk and the subsequent panel that Ray and John had been on, and they now picked right up where they'd left off, with Ray saying that the rate of improvement of technology was going to accelerate and that we were going to become robots or fuse with robots or something like that, and John countering that this couldn't happen, because the robots couldn't be conscious.

While I had heard such talk before, I had always felt sentient robots were in the realm of science fiction. But now, from someone I respected, I was hearing a strong argument that they were a near-term possibility. I was taken aback, especially given Ray's proven ability to imagine and create the future. I already knew that new technologies like genetic engineering and nanotechnology were giving us the power to remake the world, but a realistic and imminent scenario for intelligent robots surprised me.

Then he comes to the crucial point: the complexification of technology will lead to a situation where humans can no longer control the products of their own invention:

> Eventually a stage may be reached at which the decisions necessary to keep the system running will be so complex that human beings will be incapable of making them intelligently. At that stage the machines will be in effective control. People won't be able to just turn the machines off, because they will be so dependent on them that turning them off would amount to suicide.
>
> On the other hand it is possible that human control over the machines may be retained. In that case the average man may have control over certain private machines of his own, such as his car or his personal computer, but control over large systems of machines will be in the hands of a tiny elite.[6]

After reading Joy's article I decided to look into Kurzweil's opinion, the defender of transhuman technologies, and read his book *The Singularity is Near*. The book is full of interesting information about techno-scientific projects, but is hard to take seriously from a philosophical point of view. Kurzweil first ascertains that technological productivity is speeding up, and consequently, that the process of miniaturization is going to accelerate in the coming decades. He concludes that the process of creating the devices needed to replace the perishable organic tissues of the human body and of the human brain with more durable stuff will be exponential.

> The computational power to emulate the human brain approaches—we are almost there with supercomputers—the efforts to scan and sense the human brain and to build working models and simulations of it are accelerating. [...]
>
> Once the nanobot era arrives in the 2020s we will be able to observe all of the relevant features of neural performance with very high resolution from inside the brain itself. Sending billions of nanobots through its capillaries will enable us to noninvasively scan an entire working brain in real time. [...]
>
> Nanotechnology promises the tools to rebuild the physical world our bodies and brains included, molecular fragment by molecular fragment, potentially atom by atom. We are shrinking the key feature size of technology in accordance with the law of accelerating returns, at the exponential rate of approximately a factor of four per linear dimension per decade.[7]

Science fiction has already expressed the possibility of such scenarios, and I take science fiction quite seriously because I think that science fiction writers have often been the best detectors of tendencies and possibilities. But science fiction writers aren't unanimously enthusiastic or optimistic about the artificial reproduction of the human brain. Downloading human brains can lead to very horrific scenarios, as shown by cyberpunk writers.

Beyond that, several details aren't perfectly clear in the Kurzweilian promise of an eternal, happy life. I agree that the upgrading and enhancement of technical, biogenetic devices is exponentially accelerating, and that miniaturization is exponentially accelerating as well. In a few decades, we will be able to replace human bodies with non-perishable parts. I agree that this will be (and is already) a formidable improvement for medical science. Nobody can underrate such a prospect, but its philosophical implications are far from obvious.

Transhuman ideology essentially attempts to convey technical enhancement as a unified project. Kurzweil explicitly expresses this: upload the structure of a person's mind and download it into a trans-biological body. But once non-biological human intelligence is implemented, the nature of human life will be radically altered.

Understanding the methods of the human brain will help us to design similar biologically inspired machines. Another important application will be to actually interface our brains with computers, which I believe will become an increasingly intimate merger in the decades ahead. [...]

Uploading a human brain means scanning all of its salient details and then reinstating those details into a suitably powerful computational substrate. This process would capture a person's entire personality, memory, skills, and history. If we are truly capturing a particular person's mental processes, then the reinstantiated mind will need a body, since so much of our thinking is directed toward physical needs and desires.[8]

From an existential point of view, this project isn't so attractive: just imagine the infinite sadness of those decrepit minds contained in nimble, youthful-looking bodies. But in addition, Kurzweil's theory downplays a number of philosophical implications concerning the concept of the *self*.

In *The Ego Tunnel*, Thomas Metzinger asks the following questions:

> Who is the feeler of your feelings and the dreamer of your dreams? Who is the agent doing the doing, and what is the entity thinking your thoughts? Why is your conscious reality *your* conscious reality?[9]

These are questions that cannot be dismissed. Before we can download the self we have to know what the self is. According to Metzinger, the self is a false definition, a sort of illusory projection of consciousness.

> Conscious experience is like a tunnel. Modern neuroscience has demonstrated that the content of our conscious experience is not only an internal construct but also an extremely selective way of representing information. This is why it is a tunnel: What we see and hear, or what we feel and smell and taste, is only a small fraction of what actually exists out there. Our conscious model of reality is a low dimensional projection of the inconceivably richer physical reality surrounding and sustaining us. Our sensory organs are limited: They evolved for reasons of survival, not for

depicting the enormous wealth and richness of reality in all its unfathomable depth.[10]

Here, once again, we encounter the problem of individuation that Gilbert Simondon examined in depth. Because he ignores the problem, Kurzweil's proposal is philosophically quite weak, regardless of the reliability of his technological predictions.

The Register of Time and the Self

Kurzweil's crucial prediction concerns the *Singularity*, a word that he borrows from John von Neumann, who, in the 1950s, spoke of the "ever accelerating process of technology and changes in the mode of human life, which gives the appearance of approaching some essential singularity in the history of the race beyond which human affairs, as we know them, could not continue."[11]

Kurzweil's choice of the term *Singularity* with a capital *S* is, in my opinion, quite controversial. Indeed, the philosophical meaning of the term *singularity* with a lowercase *s* refers to the uniqueness and unrepeatability of an event. This is how Guattari uses the term, inscribing it in a philosophical tradition that dates back to Henri Bergson.

In *Creative Evolution*, published in Paris in 1907, Henri Bergson addressed the philosophical questions that Kurzweil evaded, in particular, the problems of mechanism and finalism in the evolution of nature and in the emergence of consciousness. Although everything in Kurzweil's predictions and theories may be right, his post-biological transhuman construct forgot something important about consciousness, which is duration, existence in time. Let us resort once more to the concept of ek-sistence. For Heidegger, ek-sistence implies the *Da*, which means presence in space, but also presence in time, being here, now. Bergson quoted Laplanche in order to define a mechanical vision that was similar to Kurzweil's vision.

> An intellect which at a given instant knew all the forces with which nature is animated, and the respective situations of the beings that compose nature—supposing the said intellect were vast enough to subject these data to analysis—would embrace in the same formula the motions of the greatest bodies in the universe and those of the slightest atom: nothing would be uncertain for it, and the future, like the past, would be present to its eyes.[12]

The problem is that the perfection of such a mechanist reduction does not grasp the situationality of existence, and therefore does not grasp consciousness and existence as singularities, with a lowercase *s*.

> The essence of mechanical explanation, in fact, is to regard the future and the past as calculable functions of the present, and thus to claim that *all is given*. On this hypothesis, past, present and future would be open at a glance to a superhuman intellect capable of making the calculation. Indeed, the scientists who have believed in the universality and perfect objectivity of mechanical explanations have, consciously or unconsciously, acted on a hypothesis of this kind.[13]

Computation can enable the perfect reproduction of a body, and also (who knows?) the perfect reproduction of a brain. But computation is missing the relation between the construct and the environment, as well as the process that Deleuze and Guattari call a *becoming other*, and that Heidegger calls the *being-towards-death*

(*Sein-zum-Tode*) of existence. Computation is missing duration, and the living perception of time.

According to Bergson:

> The universe *endures*. The more we study the nature of time, the more we shall comprehend that duration means invention, the creation of forms, the continual elaboration of the absolutely new. [14]

And also:

> Repetition is therefore possible only in the abstract: what is repeated is some aspect that our senses, and especially our intellect, have singled out from reality, just because our action, upon which all the effort of our intellect is directed, can move only among repetitions. Thus, concentrated on that which repeats, solely preoccupied in welding the same to the same, intellect turns away from the vision of time. It dislikes what is fluid, and solidifies everything it touches. We do not *think* real time. But we *live* it, because life transcends intellect. [...] *Wherever anything lives, there is, open somewhere, a register in which time is being inscribed.*[15]

Bergson is speaking of a register where life is recorded, where time is inscribed. But how can we be in touch with the register of time, how can we feel the flow of living matter? Memory is our access to this register of time, and as everyone knows, memory is not a regular, fixed, repeatable, computable reenactment of an event, or of a series of events. Memory is the re-creation and re-imagination of a past that is continuously changing as long as we distance ourselves and our viewpoint changes. Memories can be simulated, technically produced and technically inserted into a human brain, as happens to the replicant Rachel in *Blade Runner*. But living memory, the changing memory that perceives the duration of time as the marker of the self as becoming can hardly be technically reproduced.

Discrete Continuum Sensibility

Techno-transhumanism is based on the assumption that the miniaturization and extreme improvement of the digital circuits of the intelligent automaton will enable the final replacement of the frail human body with a more enduring android whose physical and mental features will be undistinguishable from the human itself.

This may be. My question is not about the technical feasibility of a perfect simulacrum, but about the emergence of the *self*, by which I mean the self-perception of a conscious singularity.

The sense of duration is the essential mark of the conscious self, and the irreducibility of existence to algorithmic recombination is based on this. The sense of duration cannot be simulated in an artificial construct that perfectly reproduces the features of the human body, because the sense of duration is not behavior, but suffering, consciousness of the organism's decomposition, and consciousness of death.

Here we are confronted with one of the fundamental philosophical questions, that is, the infinite divisibility of matter and the human impossibility to think infinity. Zeno's paradoxes, particularly the so-called *dichotomy paradox* (that which is in locomotion must arrive at the half-way stage before it arrives at the goal), deal essentially with this problem, one that, for centuries, philosophers, physicists, and mathematicians have repeatedly tried to solve.

The mathematical concepts of the discrete and the continuous are a modern reformulation of the same paradox. If matter is infinitely divisible, how is it that we experience material things as a continuum?

There isn't really an answer to this question, because existential experience and logical thinking do not belong to the same kingdom of being. While logical thinking is timeless, experience happens in time.

In *Gödel, Escher, Bach*, Douglas Hofstadter reflects on artificial intelligence and the discrete-continuous relation using mental constructions that are logically impossible, but that nevertheless exist, such as Escher's drawings.

In the introduction to the book, Hofstadter posits the problem of the relation between the consciousness and the unconsciousness of computers, and declares that the main goal of his book will be to question the unbreachable gulf between the formal and the informal.

> Computers by their very nature are the most inflexible, desireless, rule-following of beasts. Fast though they may be, they are nonetheless the epitome of unconsciousness. How, then, can intelligent behavior be programmed? Isn't this the most blatant of contradictions in terms? [...] It is not a contradiction at all. One of the major purposes of this book is to urge each reader to confront the apparent contradiction head on, to savor it, to turn it over, to take it apart, to wallow in it, so that in the end the reader might emerge with new insights into the seemingly unbreachable gulf between the formal and the informal, the animate and the inanimate, the flexible and the inflexible.[16]

The problem that Kurzweil had not elaborated is here on the table. How can we build the animate starting from the inanimate? How can we build the flexible starting from the inflexible? Zeno would tell us it is impossible, although in our daily experience we walk, and move. The jump from the discrete to the continuous is possible, although it is unthinkable. What would the quantum leap be that would enable the transformation of an assemblage of nano-devices into a sentient organism? I would say that this quantum leap, this transfer from one dimension to another dimension is sensibility, the faculty to sense one's own sensing. But can an artificial construct sense its own sensing?

> This is what Artificial Intelligence research is all about. And the strange flavor of AI work is that people try to put together long sets of rules in strict formalisms which tell inflexible machines how to be flexible. [...] The flexibility of intelligent machines comes from the enormous number of different rules, and levels of rules. The reason that so many rules on so many levels must exist is that in life a creature is faced with millions of situations of completely different types.[17]

Hofstadter's answer to the question of whether an artificial construct can sense its own sensing is obviously both yes and no. Yes in theory, since its logical impossibility can be overcome by the quantum leap of sensibility. But no in practice, because in life a creature is faced with millions of situations of completely different types, and sensibility exists in a context, and in every possible context it senses differently. Building an artifact which would be as flexible as a living and sentient organism would be an infinite task. And infinite tasks cannot be accomplished.

Although the jump from the discrete to the continuous is unthinkable, and although we are composed of discrete matter, atoms of non-living matter, we nonetheless live, think, and feel in a continuous way. Why so? It is because the dimensions of existential experience and of logical thought occur on different levels and cannot be fully translated, or made compatible. As Hofstadter writes:

> The nervous system is certainly not a discrete state machine. [...] One cannot expect to be able to mimic the behavior of the nervous system with a discrete state system.[18]

We could describe the nervous system in terms of particles, of discrete building blocks (neurons) giving rise to cognitive processes via concatenation, but this description would be less accurate than a description in terms of waves, interactions, and context-related reconfigurations. The synapse may be described as a connection, but this would not be perfectly accurate, as synapses are vibratory processes that approximate a never-achieved stability. Cognitive activity should not be defined in terms of connectivity, but in terms of vibrating conjunction. This is possibly the meaning of the expression *inner touch* through which Heller-Roazen defines consciousness.[19]

According to Diane Ackermann, awareness is essentially proprioception, the perception of ourselves in a physical, bodily sense. In my opinion, consciousness is vibratory because it is the perception of an ever-changing conjunctive concatenation: the conjunction of brain and body, but also, on the one hand, the conjunction of the body-brain and the matter entering the physical composition of the brain, and on the other, the conjunction of the body-brain and of the environment in its infinite variation.

> Consciousness is a very special phenomenon, because it is part of the world and contains it at the same time. All our data indicate

that consciousness is part of the physical universe and is an evolving biological phenomenon. Conscious experience, however, is much more than physics plus biology—more than a fantastically complex, dancing pattern of neural firing in your brain. What sets human consciousness apart from other biologically evolved phenomena is that it makes a reality appear *within itself.* It creates inwardness; the life process has become aware of itself.[20]

Essential to inwardness is time, the perception of the ongoing, irreversible dissolution of the self. Heidegger calls it *being-towards-death* and speaks of the *angst* that comes with it. This angst, which has no object, is not the perception of a danger but the constant consciousness of a process of self-consumption. Within this *angst* (which is a condition of joy no less than a condition of pain) resides the source of aesthetic desire. Aesthetic desire is not an absence, a lack of, but the vibrational emergence of a syntony, a conjunction.

9

The Horizon of Mutation

The history of Western civilization is a history of creeping abstraction. Financial capitalism means essentially submission of social life to the firm regulation of abstract calculation.

Everyday, fifty times the amount of total GDP is traded on financial markets around the world. If there is one thing we can learn from the current global economic mania, it is that capitalism cannot help itself. It operates by its own logic. It has to think short term. And so we now live in a short-term civilization.

A decade ago, Athens was the center of a global celebration of Olympic triumph. Look at it today. Heroin, the drug of depression, has taken over the streets. Cocaine, the drug of success, is moving south and east. [...]

What can be said of a civilization where oil spills can yield a profit and freestanding forests become a detriment to economic growth? What can be said of a world where the response to dwindling fish stocks is bigger nets and bigger boats? In a time of unprecedented human wealth, we now face the unprecedented threat of global ecological collapse.

We need to ask ourselves how our understanding of household management got so abstracted from reality.

—*Adbusters*, June 2013

The thread that I have been unrolling throughout this text is the mutation of sensibility in an age of creeping abstraction, and the dissonance and pain that this mutation brings about. However, I would also like to offer some keys to imagine a line of escape, starting from the investigation of the relation between aesthetic perception and production and political strategy.

Aesthetics and politics are not to be linked through an act of decision and will, as proposed by the philosophy of commitment in late modernity. They are linked because the contemporary technocultural mutation is affecting cognition, affection, and sensibility, and because art is involved in the mediascape. Consequently, it is also involved in searching for forms of autonomy from that (colonized) mediascape.

The history of twentieth-century art has expressly dealt with abstraction. At times, artists have sided with the abstraction process, at times they have tried to withdraw, at other times they have fiercely reacted, and have reclaimed the body, eroticism, irregularity and dreams as possible antidotes to the cold poison of abstraction.

Semiocapitalism captured and exploited the energy coming from sensibility and from rebellious art, but simultaneously transferred this energy into the abstract dimension of the market, of design, and of virtual technology.

In the last chapter of this text, I want to envisage what the next game will be. Many signs suggest that the game of progressive humanism is over, and that connective abstraction has recoded language, frustrated any attempt at resistance and any expectations of a possible reversal of the current trend.

As I have argued in the earlier sections of this text, a mutation is underway, a mutation that can be neither resisted nor reversed. Resisting a mutation, or refusing to acknowledge the effects of a mutation implies self-marginalization and the inability to understand the current transformation and interact with those who are affected by it.

> No age lets itself be done away with by a negating decree. Negation merely throws the negator off the track. Modernity requires, however, in order, in the future, for it to be resisted in its essence and on the strength of that essence, an originality and breadth of reflection for which, perhaps, we moderns can prepare somewhat, but over which we can certainly never gain mastery.[1]

The mutation must be interpreted, so that the conscious and sensible organism may find new ways toward autonomy and self-deployment in the fields of aesthetics, ethics, and politics.

Reviving the intensity of bodily sensibility, and disentangling the potency of the general intellect from the techno-economic apparatus are the cultural and political tasks of the future, and they are narrowly linked.

How can we successfully accomplish these tasks? How can we respond to this prospect?

On the last page of the Summer 2013 issue of *Adbusters* is a beautiful image with a text by Oscar Niemeyer that reads:

> I am not attracted to the right angles or to the straight line, hard and inflexible, created by man. I am attracted to free-flowing, sensual curves. The curves that I find in the mountains of my country, in the sinuousness of its rivers, in the waves of the ocean and on the body of the beloved woman.

Error, dissonance, excess provide the starting point. But we must still understand what the landscape of the next game will be; we must still imagine what the horizon of the next evolutionary step will be. Let me try to outline a methodology for disentanglement.

Then We're Stupid, and We'll Die

During the last fifteen years, a large social movement has questioned capitalist absolutism: protests, demonstrations, and political actions have tried to stop the devastation of the social civilization that we inherited from the past.

Since November 1999, when a hundred thousand activists and workers disrupted the World Trade Organization summit, a worldwide movement of protest has spread, notwithstanding the violent reaction on the part of the armed forces of nation states (the nation state that has lost all function and sovereignty, except to repress social unrest). After the financial collapse of 2008, a new wave of movements against capitalist absolutism spread throughout Europe, the United States, and elsewhere.

Regardless of their dimension, social extension, and moral prestige, these movements have been totally ineffective. Not one devastating corporate project has been stopped, not one cut in social welfare has been prevented.

Why so?

I see two reasons for the current impotence of social movements.

Firstly, movements are strong in the streets but unable to attack the economic interests of corporations because the precarization of labor has destroyed solidarity at the level of production, and solidarity is the only material force that can oppose the material force of corporate interests.

Secondly, the abstract feature of financial capitalism is unattainable through the concrete forms of social action.

What then?

The effects of capitalist absolutism are already apparent in the form of a multiple catastrophe; in the environment, in social welfare, and in education, unimaginable havoc and unspeakable suffering are already underway.

"Then we're stupid, and we'll die," says Pris, one of the four androids in *Blade Runner* who escaped the extra-terrestrial colony and came to earth to meet the only person who could extend her life, the God of Biomechanics, the CEO of the Tyrell Corporation who created the body and brain of the Nexus generation.

In the incipit of his article "Accelerationist Aesthetics: Necessary Inefficiency in Times of Real Subsumption," Steven Shaviro quotes Mallarmé:

> *Tout se résume dans L'Esthétique et l'Economie politique.* Everything comes down to Aesthetics and Political Economy. Mallarme's aphorism is my starting point for considering accelerationist aesthetics. I think that aesthetics exists in a special relationship to political economy, precisely because aesthetics is the one thing that cannot be reduced to political economy.[2]

This is an interesting point, and I want to explore it.

Aesthetics converge and collide with the economy. As long as the social body is unable to get rid of the process of ever-expanding abstraction, aesthetic research will share a border with psychopathology, and will concern stress, acceleration, and suffering.

Earlier in this book, I charted the aesthetic genealogy of the economy.

I will now try to scrutinize the other side of the problem, that is, the incompatibility between the economy and aesthetics. While the current economy is about abstraction, aesthetics refers to sensitive knowledge. While the economy is about abstraction, aesthetics is about the concrete experience of the sensitive mind. This is why it offers a line of escape (but not of resistance).

Experience in Technomaya[3]

A famous ascetic named Nārada, having obtained grace of Vishnu by his numberless austerities, the god appears to him and promises to do for him anything he may wish. "Show me the magical power of thy *māyā*," Nārada requests of him. Vishnu consents, and gives the sign to follow him. Presently, they find themselves upon a desert road in hot sunshine, and Vishnu, feeling thirsty, asks Nārada to go on a few hundred yards farther, where there is a little village, and fetch him some water. Nārada hastens forward and knocks at the door of the first house he comes to. A very beautiful girl opens the door; the ascetic gazes upon her at length and forgets why he has come. He enters the house, and the parents of the girl receive him with the respect due to a saint. Time passes, Nārada marries the girl, and learns to know the joys of marriage and the hardships of a peasant life. Twelve years go by; Nārada now has three children and after his father-in-law's death, becomes the owner of the farm. But in the course of the twelfth year, torrential rains inundate the region. In one night the cattle are drowned and the house collapses. Supporting his wife with one hand, holding two of his children with the other and carrying the smallest on his shoulder, Nārada struggles through the waters. But the burden is too great for him; he slips, the little one falls into the water; Nārada lets go of the other two children to recover him, but too late; the torrent has carried him away. Whilst he is looking for the little one, the waters engulf the two others, and, shortly afterwards, his wife. Nārada himself falls, and the flood bears him away unconscious, like a log of wood. When, stranded upon a rock, he comes to himself and remembers his misfortunes, he burst into tears. But suddenly he hears a familiar voice: "My child! Where is the water you were going to bring me? I have been waiting for you more than half an hour!"

Nārada turns his head and looks: instead of the all-destroying flood, he sees the desert landscape, dazzling in the sunlight. And the god asks him: Now do you understand the secret of my *māyā*?[4]

We live in the multilayered dimension of technomaya. Digital technology has given the media the power to act directly upon the mind, so that the mediasphere casts a spell that envelops the psychosphere.

Technomaya captures the flows that proceed from the activity of the mind, sending them back to the mental receptors as a mirror, a template for future imagination, a cage for future action, and for future forms of life.

In the digital sphere, people spend more and more time with electronic ghosts. But the techno-media spell (technomaya) breaks down from time by time because its barred windows are suddenly opened by the wind of joy, or the storm of despair, so that the dazzling light of uncanny dimensions bursts on the scene of the social imagination, allowing forgotten fragments of the unconscious to surface.

I want to describe the spell of semiocapitalism—financial abstraction, mediascape ghosts—that captures the social body and delivers it over to the economic code, where experience is subjected to the power of simulation and standardization.

But I also want to search for and imagine possible lines of escape. Such lines of escape can only be found in those places of the unconscious where the multi-layered spell of semiocapital is ripped apart, so that a multi-layered unconscious resurfaces.

Reality is the point of intersection of countless projections proceeding from the intentionality of living and sentient beings.

Experience is the way to access reality, but it is also the act of projecting reality onto the screen of shared perception. It is attention, but also intention.

Experience means to open one's eyes and see the existing world, but it also means māyā, the projection of a world.

The etymology of the word experience has something to do with the act of going through, *per-ire*, which also means *to die*. We also can say that *ex-perire* means to try out.

Only by passing through the tests and places that life presents does one becomes an expert. Experience is the process of living through something that you did not know before, so that you can find its singular meaning. Singularity is an essential feature of experience, since experience is the act of personalizing and singularizing a place that one knows from going through it.

> Experience is not a matter of having actually swum the Hellespont, or danced with the dervishes, or slept in a doss-house. It is a matter of sensibility and intuition, of seeing and hearing the significant things, of paying attention at the right moments, of understanding and coordinating. Experience is not what happens to a man; it is what a man does with what happens to him.[5]

Experience is not simply the act of exposing the skin and mind to the flow of stimulations coming from the environment. It also implies adapting the mind and skin to the environment, and the active projection of the expectations of the experimenter. In this sense, experience is the singularization of the environment, the singular shaping of the world.

Experience thus implies not only attentive perception, but also intentionality.

Merleau-Ponty speaks of intentionality as an act of identification that comes from the living world of the subject and projects on the world, the ob-ject that is thrown there in the outside:

'All consciousness is consciousness of something'; there is nothing new in that. Kant showed, in the *Refutation of Idealism*, that inner perception is impossible without outer perception, that the world, as a collection of connected phenomena, is anticipated in the consciousness of my unity, and is the means whereby I come into being as a consciousness. What distinguishes intentionality from the Kantian relation to a possible object is that the unity of the world, before being posited by knowledge in a specific act of identification, is 'lived' as ready-made or already there.[6]

And also:

The world is not what I think, but what I live through.[7]

And Husserl, in *Experience and Judgment*, writes:

The retrogression to the world of experience is a *retrogression to the "life-world"* (*Lebenswelt*), i.e. the world in which we are always already living, the world which furnishes the ground for all cognitive performance and all scientific determination.[8]

And also:

> Experience in the first and most pregnant sense is accordingly
> defined as a direct relation to the individual.[9]

The very concept of experience has to be re-examined in the light of the current techno-mutation.

First of all, the digital format of experience with its increasing speed and intensity affects the psychic reaction to information stimuli, the empathic relation between conscious and sensitive organisms, and cognition, that is, memory, imagination, and language. Experience as attention and intention is subject to an intense stress that results in a mutation of the cognitive organism.

Capturing Attention

The fundamental contradiction of semiocapitalism is the incompatibility of cyberspace with cybertime. Since it is the product of countless sources of virtual projection, the expansion of cyberspace is boundless.

Cybertime, on the contrary, is not infinitely stretchable. It is composed of the time of attention, which cannot be intensified beyond a certain point because of its physical, emotional, and cultural limitations.

In economic terms, the output of semio-production is by far outpacing the market of attention, which means that the phenomenon of cyclical crises that Marx described as an effect of overproduction in the sphere of industrial capitalism is no longer cyclical but permanent.

According to Jonathan Crary, author of *24/7: Late Capitalism and the Ends of Sleep*, the capitalist need for expanding markets results in a restless stimulation of social attention aimed at increasing the time of alertness.

> This is the form of contemporary progress—the relentless capture and control of time and experience.[10]

This unceasing assault on attention is causing a contraction of the time available for the emotional elaboration of information stimulation and for the rational decision-making that was the condition of politics.

This is why, so often, political choices seem devoid of rationality, and social relations become brutal and aggressive: because the time for rational and emotional elaboration is so intensely reduced that society seems to act in a whirlwind—as happens to those who sleep too little and who take drugs to stay awake.

Crary's book focuses mostly on the reduction of sleep time as an effect of the economic assault on the time of attention.

> It should be no surprise that there is an erosion of sleep now everywhere [...]. Over the course of the twentieth century there were steady inroads made against the time of sleep—the average North American adult now sleeps approximately six and a half hours a night, an erosion from eight hours a generation ago, and (hard as it is to believe) from ten hours in the early twentieth century.[11]

Sleep, in fact, can be considered an "uncompromising interruption of the theft of time from us by capitalism."[12] A society of insomniacs is not at all a comforting place, and the increase in productivity is paid for with a loss of rationality and a loss of respect for life. The irrational exuberance of financial agents who take drugs in order to trade day and night on their computers has already brought the world to the brink of an abyss, and will do so again and again.

As a conclusion, Crary suggests that "sleep is the only remaining barrier, the only enduring 'natural condition' that capitalism cannot eliminate."[13]

Let me revise that statement. There is another enduring natural condition that capitalism cannot eliminate, another enduring natural barrier to the intrusive financial hubris, that is death. Suicide is spreading everywhere, as an effect of social stress, emotional impoverishment, and the constant aggression on attention.

According to the *World Health Organization*, in the last 45 years suicide rates have increased by 60% worldwide. These were the years of the full implementation of global capitalism, years, as well, of the thorough submission of attention time to the rhythm of the economic machine.

These figures concerning suicide do not include suicide attempts, which are up to 20 times more frequent than successful suicides.

An epidemic of unhappiness is spreading on the planet while capitalist absolutism is asserting its right to the unfettered control of our lives.

Semiocapitalism is infiltrating the nervous cells of conscious organisms, inoculating them with a thanato-political rationale, a morbid sentiment which permeates the collective unconscious, culture, and sensibility—an obvious effect of sleep deprivation and a patent consequence of the stress placed upon attention.

The Google Empire has essentially been built on the capture of the experience of its users, in order to increase value and productivity. In the course of creating the most refined attention-draining machine, the personal computer has been bypassed by the release of the latest generation of cellular phones, labeled smartphones, whereby access to the network has gone mobile, pervading every moment of day and night. *Mobilizing* access to the net has obviously expanded the amount of attention-time that is captured, and subjected new dimensions of personal life to the all-pervading search for semio-profits.

> Ubiquitous computing in the era of the first smartphones focused on seamless integration, that is, naturalization of socio-technical factors in workplace settings that transcended geography and time, heralding "unprecedented levels" of productivity as well as the inevitable "interaction overload."[14]

The Map Captures the Territory: Coding Orientation

According to Jean Piaget, the process of discovery by which we become familiar with our environment and interiorize the space around us starts in the first days of life and determines the construction of our internal space, which becomes the map of how we orient ourselves and the condition of further acquisitions.[15]

Orientation is the cognitive ability to recognize the physical features of the environment around us, and to build an inner map that makes it possible to move with intention in the world. The process of internal mapping that precedes orientation implies a highly singular relation with the environment: the sensory elaboration and emotional selection of places, signs, but also lights, textures, and scents. Orientation can be understood as the way that we singularize the landscape, the process through which we make *the* world *our own world*.

Orientation is the intimate mapping of the space that we navigate and which we inhabit. The territory through which we move stimulates emotional effects in our mind: the memory of the places that we cross is the emotional marking, and therefore the singularization, of external space. Once we have recorded the points that mark a territory, this becomes our intimate map, the condition for further orientation, for new discoveries, new recordings, and for a never-ending remapping of the world.

The map of the city becomes the representation of the person who has been dwelling there, as Borges suggests in a poem where he speaks of his becoming blind, and remembers the city as the intimate map of his life.

> I live among vague, luminous shapes
> that are not darkness yet.
> Buenos Aires,
> whose edges disintegrated
> into the endless plain,
> has gone back to being the Recoleta, the Retiro,
> the nondescript streets of the Once,
> and the rickety old houses
> we still call the South.
> [...]
> Those paths were echoes and footsteps,
> women, men, death-throes, resurrections,
> days and nights,
> dreams and half-wakeful dreams,
> every inmost moment of yesterday
> and all the yesterdays of the world [...].[16]

In his book about getting lost, the Italian architect and anthropologist Franco La Cecla writes:

The word *orientation* has a double meaning, the first is essentially active, it is the ability to organize one's surroundings, to create a general frame of reference to which knowledge can be tied. The second is the passive ability to follow indications, to read a map, use a compass, adapt to a system of pre-existing coordinates in order to locate a place or to reach a destination.[17]

What I said about experience in general can be said about orientation: it is both the ability to follow cartographic instructions and the ability to draw a map.

The experience of orientation consists in getting lost in the territory that is being encountered, and in finding one's bearing again, creating a singular perception of space.

La Cecla also examines the disorienting effects that the modern standardization of urban space has induced in city dwellers.

Modern functionalism is based on the assumption that city dwellers should not waste time in a complicated relation with the environment. The environment must be functional, so that the inhabitants of the city can move from one suburban area to another in order to do their job. This functionalist reduction implies the anonymity of the suburbs: any emotional transfer is considered useless. The environment must not be felt, but used. This transformation leads to the standardization of orientation.[18]

The creation of those functional places, which Marc Augé has labeled *non-places*, leads to the obliteration of the singular relation between the individual's mind-eye-body and the space around it. Little by little, the modern reshaping of territory, aimed at increasing the productivity of urban territory and at facilitating car transportation in metropolitan areas, has erased the marks of the historical past, and, more generally, the signs that have been ingrained with emotional memories. In a text written at the beginning of the twentieth century, Rainer Maria Rilke identified the standardization of places with the influence of America.

> Even for our grandparents a "house," a "well," a familiar tower, their very clothes, their coat: were infinitely more, infinitely more intimate; almost everything a vessel in which they found the human and added to the store of the human. Now, from America, empty indifferent things are pouring across, sham things, *dummy life*.[19]

Although modern architecture and urban design have contributed to the standardization of territory, the spread of the digital has paved the way to a further shift towards the ultimate de-singularization of orientation. The map is capturing the territory, and, as more and more people use tools for guided navigation, based on the geo-tagging technology of Global Positioning Systems (GPS), orientation is turned into mere functional navigation. Smartphones are used to access a geo-referential cartography, and lead to the devolution of orientation into the mere ability to interface with an interactive digital map.

In 1973, the first attempts at developing a global positioning system were instigated by the American military under the name of the *Defense Navigation Satellite System* (DNSS), later coined the *Global Positioning System*. A GPS receiver calculates its position through a precise timing of signals sent by GPS satellites high above the Earth. Each satellite continually transmits messages that include the time the message was transmitted and the position of the satellite at that time. Using the speed of light, the GPS receiver computes the distance to the satellite, narrowing the receiver's position to the surface of a sphere. By locking-in to four different satellites, a 3-dimensional position can be calculated. This location is then displayed within a moving map.

With this in mind, it's easy to predict a rapid atrophy of the sense of orientation as it is replaced by the technology of positioning. Over the course of the next generation, the mental processes that consist in the internal mapping of territory might devolve by reliance upon geotagging machines, and the ability to identify our place in the world and to singularize our surrounding landscape might fade, and perhaps nearly disappear from our connective mind.

Together with the experience of getting lost, the experience of recognizing a place will fade or, at least, grow quite dull. We can understand this fading of the faculty of orientation as one step in the process of the connective reshaping of experience as a whole.

Swarm Experience

In the meantime, a new device has appeared—a new, wearable interface between the mind and the world that represents a new step in the cognitive mutation underway. Google, the most revolutionary corporation, and the most perfect colonizer of all time, has paved the way to the ultimate obliteration of singularized experience, and therefore to the cancellation of singularized processes of living in the world (*Lebenswelt*).

During the first decade of this century, Google acted as the universal draining pump of meaning. Capturing and collecting billions and billions of individual acts of meaning-allocation from countless users worldwide, Google has created the most flexible machine of de-singularization ever conceived of. The results of a Google search are influenced by its user's previous search queries and results. Google knows what you need, and you know that Google knows what you know, and Google knows that you know that Google knows what you know. So the results to your queries will exactly coincide with your needs: Google continually refines your queries just as the search engine itself becomes more precisely customized to your needs. The end result is that the user's world is tailored by the feedback system of queries and results of the Google-user interface.

As I'm writing these pages, in the year 2013, the most flexible, most free—and most totalitarian—corporation is launching GoogleGlass, a product that promises to be the ultimate coder— and decoder—of the human experience.

In the not-so distant future, users wearing GoogleGlass will receive information about their objects of vision directly on the screen before their eyes, in the space between themselves and those objects. In other words, GoogleGlass will be a wearable computer with an optical interface displaying information about objects in the user's field of vision.

Let's suppose you are in Rome, standing in front of the Coliseum. You click on your GoogleGlass and receive information about the monument in front of you. As the information that Google makes available is composed of the average information uploaded by Google users, the experience of the GoogleGlass user will grow increasingly more uniform.

Let's suppose you meet a stranger: GoogleGlass will tell you who this person is, so you can interact with him or her according to the suggestions and implications that you'll have been led to draw from GoogleGlass information.

Little by little the entire world—already entirely mapped by Google maps—will be re-coded by GoogleGlass, so you can access those previously undergone experiences that GoogleGlass makes available for you.

This implies that you will no longer experience the world, but, rather, that you will simply use (or receive, or access) previously-experienced data about an object that is no longer the object of your own experience but purely a reference to a pre-packaged world.

As reality is the point of intersection of our projections, and experience is our singular access to the world of life and to the creation of shared meaning, this techno-mutation will come to affect reality itself.

The world, as experience and projection, will be evacuated and replaced by a uniformed, simulated experience—the experience of the swarm.

Neuro-Totalitarianism in the Making

According to Giovanni Gentile, the philosopher of Italian fascism, totalitarianism is a political regime in which everything—from the economy, to the educational system, and to ethical behavior—is subjected to the action of the state. The process of connective creation of the swarm has nothing to do with the fascist form of totalitarianism, but it can be described as a process of standardization of cognition, perception, and behavior based on the inscription of techno-linguistic automatisms in human communication, and therefore in the connective mind.

This form of techno-totalitarianism results from three consecutive steps.

The first step is the permanent connective wiring of the inter-actions between humans.

The process of *cellularization* has been the perfect carrier of this socio-cognitive mutation. More pervasive than the computer, the cell phone has finally created the infrastructure of global intercon-nection and is paving the way to the ultimate deterritorialization and ubiquity of information.

The social effect of this process of deterritorialization and con-nection, which has already been widely implemented, can be seen simultaneously in the globalization of the labor market and the precarization of work, but also, paradoxically, in the utter indi-vidualization and the inescapable collectivization of personal lives. Neoliberal ideology emphasizes individualism, but the competitive consumerist individual is extremely standardized in his or her goals, tastes, and desires. Individualism and singularity have little in common. Contrary to individualism, singularity is not competitive, exchangeable, or standardized.

Cellularization has accomplished a process that Habermas described as "the uncoupling of system and lifeworld"—the separation of living language (the voice, the singularity of the speech act) and the perfection of a techno-linguistic system of permanent exchange between speakers who are less and less actors of their own interaction, and increasingly acted upon by techno-linguistic interaction.[20]

Cellularization—i.e. the connection of every agent of enuncia-tion in the network—is the general framework of the subsumption (or capture) of social communication into the electronic swarm.

Therefore cellularization is the full implementation of what Heidegger calls the language of technology, implying that technology is the subject of language, and that language is spoken by the tech-nological system.

The second step in the process of instituting neuro-totalitarianism is the current replacement of living experience, and its simulation with standardized, recorded stimulations, a process that I have analyzed above in terms of the automation of the sense of orientation.

These first and second steps are concerned with cognitive activity and its psychological implications—the software of the mind.

The third step toward the implementation of the swarm is directly aimed at modifying the neural hardware itself: namely, the insertion of techno-devices, nano-prostheses, modifiers and enhancers of neural programming in neurological systems. Such manipulation of neural systems is not a new phenomenon; psychopharmacology is already acting on neural matter, particularly on the neurotransmitters that regulate mood, attention, and the reactivity of the psyche. But we should expect more neural manipulation in the future. In April 2013, the President of the United States declared that one of the most important American investments in the field of scientific research was to be *Brain Activity Mapping*, also known as *Brain Research through Advancing Innovative Neurotechnologies*, which intends to map the activity and functions of every neuron in the human brain.

This project is based on the assumption of neuroplasticity, i.e. the possibility of intervening on the neural system, redirecting neural activity, and reshaping synaptic pathways.

Neuroplasticity, however, is an ambiguous condition that provides the opportunity of an alternative. In fact, the possibility of transforming the processes and material structures of cognitive activity, of reshaping synaptic pathways, although it opens the way to the neuro-totalitarian domination of semio-corporations (the media) and psycho-corporations (psychopharmacology), nonetheless also invites a process of sabotage and subversion of the dominant mode of mental wiring, opening the way to experimenting with forms of free neuro-psychic concatenation that correspond to the social processes of self-organizing cognitive work.

Resisting Mutation?

Should we plan to resist the mutation underway?

That would be a reactionary and technophobic choice, and, moreover, an impossible task. I don't think that resisting mutation is possible.

Technological innovation generates tools that reshape our social environment, empowering individuals who adhere to it and cutting off those who resist it. This is why resistance would be futile. Individuals cannot resist the capture that occurs when change happens in the field of communication devices. And further, the diffusion of network technologies hastens the pace of integration.

Accepting the challenge is, therefore, unavoidable, and it is only in accepting this challenge that we can see a possible alternative for tomorrow.

The alternative that we must face in the future is now apparent: it is the choice between the totalitarian submission of the nervous system to the semio-financial governance of capitalism—and disentangling nervous energy (and of the activity of the general intellect which is the organized expression of nervous energy) from the semio-financial rules embedded in the governance of the system.

This will be the major game to be played out in the coming decades.

All other power games have been played, and all of them have been lost.

But the game of neuroplasticity is only beginning. It aims to disentangle the autonomy of the general intellect from its neuro-totalitarian jail, which corresponds to the needs of absolute capitalism. At this point, we can now only glimpse the possibility of disentangling mental activity from the spell of technomaya.

The general intellect will either be coded by the semiotic matrix of the semio-economy and social activity turned into a swarm connected at the techno-neural level, or the general intellect will reunite with its sensible body to create the conditions for the independence of knowledge from the matrix and for the singularity of experience.

Morphogenetic Vibration in *Neuromagma*

How does a new form emerge from the magma of matter and information? This question is crucial if we want to imagine the possibility of disentangling social potency and the general intellect from the double bind of capitalism.

The emergence of a new semiotic form occurs in the space between neuromagma and vibratory morphogenesis.

Capitalism is essentially a semiotic framework, it is the form that semiotizes (codifies) contents according to certain interests, and to the paradigm of accumulation.

The entangling form (capital) is the semiotic codifier that bends contents (social skills, collective intelligence) and perverts their potential into repeating the economic model of accumulation.

The continuous process of transformation of the global mind can be described as magma, a chaotic ebullience of inter-individual synaptic pathways: conjunctions, and the sudden proliferation of neurons that escape existing connective patterns.

Neuromagma contains infinite possibilities of evolution, yet existing forms of neural wiring are forcing neuronal pathways into pre-packaged connective circuits. How can we disentangle diverging possibilities of conjunction for the neural activity of the global mind?

Morphogenesis (the creation and emergence of new forms) comes from the vibration of neuromagma. This vibration exceeds wired connectivity to the point of rupturing (disconnecting) the existing circuit.

Forms emerge from the interaction between the internal structure of connectivity and the external environment of neuromagma. Neuromagmatic machines can disrupt connective structures, and give rise to a vibratory dynamics in search of new semiotizations, new forms.

The vibratory interaction of a system with its environment marks the beginning of a new form. In an article entitled "On the Origin and Nature of Neurogeometry," Alessandro Sarti and Giovanna Citti try to understand the problem of individuation from the point of view of neurogeometry, and particularly from the point of view of the neurophysiology of the vision process. They explain that visual cells activated by stimulus-images enter in communication through horizontal connectivity.

Sarti and Citti write:

> The plasticity of the brain, i.e. its ability to reorganize neural pathways based on new experiences through learning procedures, guarantees a strong connection between the design of our perceptual systems and the properties of the physical environment in which we live.[21]

In conditions of extended criticality and vast complexity—billions of cells propagating and colliding, or billions of human beings following different projects, entering into conflicts, and struggling for life—a morphostatic topology explodes. We can therefore say that inert—quiescent, motionless—geometry enters a vibratory dynamics.

Chaosmosis is the process that follows the explosion of a morphostatic topology, and the resulting emergence of a new form.

How does a collective territory, a crowd, a society, or a network, enter into vibration? It happens through resonation or resonance. We can try to understand resonance in terms of the relation between a rhythm and a refrain. A rhythm is the relation of a subjective flow of signs, musical, poetic, or gestural, with the cosmic, earthly, or social environment. A rhythm is singular and collective. It singularizes the sound of the world in a special modeling of environmental sound. But it can trigger a process of agglutination, of sensitive, and sensible communality. For example, when people start to sing the same song, and dance the same dance. This can be dangerous when the sense of community is based on the arbitrary presumption of a natural belonging. Fascist subjectivation was, for example, based on such mandatory homogeneity. But this can also occur in ironic and nomadic ways. People start to create a new song, and do it together. They don't do it because they share a common identity, or because they think they belong to the same territory or destiny. They do it because they like the sound of their voices vibrating in unison. This is what I call a movement. What is a movement? It is an event that opens a new landscape. In the field of art or social politics for example, when a movement occurs (literally, a displacement) we become able to see things that we did not see before.

A refrain is an obsessive ritual that allows a singularity—a conscious organism in continuous variation—to find points of self-identification, and to territorialize and perceive itself in relation to the surrounding world. A refrain is a mode of semiotization that allows a singularity (a group, a people, a nation, a sub-culture, or a movement) to receive and project the world according to inter-individual, that is, reproducible and communicable formats. In order for the cosmic, social, and molecular universe to be filtered through individual perception, the individual mind activates filters or models of semiotization. I call these filters refrains. The way a society, a culture, or a person perceives time is also a model of a truly temporal refrain, that is, of particular rhythmic modulations that act as ways to access cosmic temporal becomings and to attune to them.

From this perspective, universal time appears to be no more than a hypothetical projection, a time of generalized equivalence, a *flattened* capitalistic time; what is important are these partial modules of temporalization, operating in diverse domains (biological, ethological, socio-cultural, machinic, cosmic...), and out of which complex refrains constitute highly relative existential synchronies.[22]

The stuff that composes a refrain is essentially rhythm. Singular refrains can create a common space of resonance, and a new form emerges as a new rhythm. It is the new rhythm that makes it possible to see a new landscape.

Art, Sensibility and Neuroplasticity

The process of transformation, which was the object of political imagination in modernity, is shifting to the conceptual and practical sphere of neuroplasticity.

The mutation of the mind is underway. It is the consequence of a spasmodic attempt by individual minds to cope with a chaotic global infosphere, and to reframe the relation between the psychosphere and the infosphere, between cognition and stress, and between the brain and chaos.

Phenomena of traumatic adaptation are moving through the space of the social brain. Not only is the psychological dimension of the unconscious disturbed, but the fabric of the neural system itself is also subjected to trauma, overload, and disconnection. The brain's adaptation to its new environmental conditions involves enormous suffering, a tempest of violence and of madness.

A wide range of contemporary pathologies seem to escape psychoanalytical frameworks, and concern *cerebrality* rather than sexuality—as Catherine Malabou puts it in *The New Wounded: From Neurosis to Brain Damage*.

Alzheimer's, post-traumatic stress disorder, panic, attention deficit disorders, autism and anorexia are symptoms of a disturbance that is affecting not only linguistic and psychic software but neurological hardware as well.

The problem is, does consciousness play a role in this process of mutation? Does imagination consciously act on neuroplastic processes? Can the conscious organism do something when it is taken up in a situation of spasm?

In Guattari's parlance, chaosmosis is the overcoming of the spasm, the relaxing of spasmodic vibration. Chaosmosis is the creation of a new, more complex order (syntony and sympathy) out of the chaos created by the spasmodic acceleration of the surrounding semio-universe. Chaosmosis is the osmotic passage from a state of chaos to a new order. Here, the word *order* does not imply any normative intention, and has no ontological meaning. It is the harmonic relation between the mind and its semio-environment, sympathy, and shared perception.

...people are constantly putting up an umbrella that shelters them and on the underside of which they draw a firmament and write their conventions and opinions. But poets, artists, make a slit in the umbrella, they tear open the firmament itself, to let in a bit of free and windy chaos and to frame in a sudden light a vision that appears through the rent—Woodsworth's spring or Cezanne's apple [...]. Art is not chaos but a composition of chaos that yields the vision or sensation, so that it constitutes, as Joyce says, a chaosmos, a composed chaos—neither foreseen nor preconceived.[23]

Poetry is the linguistic chaoide that reopens the space of indetermination, re-establishing the autonomy of enunciation from the functioning of techno-linguistic interfaces.

Poetry is the ironic act of exceeding the established meaning of words.

In every sphere of human action, a grammar establishes limits that define a space of communication.

Today, the economy has become the universal grammar penetrating every level of human activity. Language is defined and limited by its economic exchangeability.

The reduction of language to information and the incorporation of techno-linguistic automatisms into the social circulation of language are securing the subjection of language to financial economy. However, whereas social communication is a limited process, language is boundless, its potentiality goes beyond the limits of the signified. Poetry is the excess of language, the signifier disentangled from the limits of the signified. Irony, the ethical form of the power of language to exceed, is the infinite game words play to skip established meanings, to shuffle them around, and to create new semantic concatenations.

Social movements, at the end of the day, can be viewed as ironic acts of language, as semiotic insolvencies, refusals to pay the debt of meaning, and ultimately as the disentanglement of language, behavior, and action from the limits of symbolic debt.

The conscious and sensitive organism evolves through the inter-action of individual and collective spheres, through the link between individual neuro-activity and connective concatenation. The neuro-plasticity of an organism's sub-individual components (the molecular decomposition and recomposition of biological matter) interacts with the rhythms and the super-individual automatisms of the techno-linguistic swarm, the bio-informational super-organism that is embedded in the totalitarian governance of semiocapitalism.

Techno-linguistic interfaces link the organism to the bio-info super-organism of the Net; language is subjected to automated wiring. Cognition is caught up in the inescapable loop of this endless self-confirmation.

Only an excess of imagination can find the way to a conscious and consciously managed neuroplasticity, but we cannot know if imagination-excess can still function when its cognitive wiring is set.

This is the question that we will have to deal with in the coming decades. It is the next game, the neo-human game that we can still only barely sense beyond the apparently unstoppable and irre-versible catastrophe of human civilization that is underway.

10

Consciousness and Evolution

Determinism as Description and Strategy

Determinism is a philosophical theory based on the assumption that every physical event can be reduced to a chain of causations. Over the course of the last century, deterministic vision simultaneously gained and lost ground.

It gained ground thanks to the refinement of tools for microphysical investigation, and the discovery of highly complicated patterns of causation and determination in the field of biogenomics. But it lost ground both in physics and in biology since Heisenberg's principle of indetermination blurred the link between physical phenomena and the observer.

In the last decade of the century, the Genome Project was founded on the presumption that genes determine an organism's life in the same way that a code enables the interpretation of a message. Such a deterministic reduction was particularly important for the conceptualization of the Genome Project, following the idea that an organism's deployment is contained in the informational structure of its genetic code. Once the genome was mapped, however, scientists and philosophers dispelled the deterministic implications of biogenetic mapping by stressing the importance of epigenesis in the formation of living organisms. Epigenesis is the aleatory and unpredictable effect that the environment exerts on the unfolding of the organism's bio-info code, as it transits from zero-dimensional information to a multi-dimensional body.

According to epigenetic theory, code is indeed only the palette (or range of possibility) in which the generation process occurs by selecting one possibility among many. A living organism must not be understood as the predictable unfolding of information contained in its code. Rather, it should be seen as the result of an infinitely complex negotiation between the code and the environment, between the possibilities contained in the code and the outcome of epigenetic deployment. Epigenesis is the vibratory oscillation of a genetic flow of information that meets the environment, and changes direction and shape according to environmental events.

Epigenetic theory reveals the flaw of determinism. Based on the assumption that phenomena are the predictable deployment of a coded process of generation, determinism misses the aleatory process of vibration that leads to the implementation of one possibility among many. At the end of the day, this is why determinism must be abandoned as a methodology for describing reality. But determinism is not simply a description, it is also a project, and from this point of view we should look at it in a new way. Indeed, determinism can be viewed as the strategy of inserting deterministic tools into the living organism and its brain, cognitive automation being the technology to inject determinism into the human sphere.

Cognitive Automation

The automation of cognitive activity will be a major trend in the coming age, and marks a leap into a post-historical dimension. In the modern, humanist sense, history was the process of the conscious affirmation of freedom projects in the field of political action. But the cognitive mutation we are speaking of will dissolve the historical relation between consciousness, politics, and freedom. Automation is replacing political decision. The word *governance* essentially refers to this automation of decision-making and of the interpretation of data, it implies the end of politics, democracy and conscious decision-making, and the establishment of an automatic chain of logical procedures intended to replace conscious voluntary choices. It is transforming the social organism into a swarm.

In current conditions of hyper-complexity—the high intensity and excessive speed of info-stimulation affecting the brain—social action is less and less the result of conscious organized choices, and increasingly the result of automated chains of cognitive elaboration and social interaction.

In the current (neo-human) transformation based on the digital manipulation of language and life, a kind of theological dependence of human action is resurfacing. Techno-linguistic automatisms are in fact acting like a post-humanist God, whose operational force is inscrutable and superior to human action and will.

It is not surprising that the neo-human is emerging within American cultural space. Since the beginning of colonization, North American civilization was marked by the erasure of a cultural past. The cultural and religious legacy of the European past was cancelled by the Puritan decision to abandon the old continent, which had been polluted by religious and political corruption. The elimination of indigenous populations (the most perfect genocide ever) cleared American space from any trace of local culture. In a territory purified from the historical and cultural traces of the past, the Puritans built a new civilization that was essentially based on a verbal relation with God.

The Puritan language works through yes or no. There are no nuances or ambiguity, only the perfect alternative: zero or one. In their relation with God, Puritans did not rely on images, on impure, ill-defined representations. Words alone made the relation between God and the elected people possible.

Catholics invaded the Mexican cultural landscape with a flow of imaginary, and based the process of evangelization on the ambiguity of images whose interpretation was infinite and constantly open to semantic negotiation. The Anglo-Saxon Puritans, on the contrary, laid the foundations for a binary infosphere that contained no ambiguity, where every question could have only one answer: yes or no.

This is the epistemic and aesthetic condition of the neo-human civilization that led Western culture to create computing technology and to digitalize the infosphere, and that is now heading toward the utopia of the final automation of intelligent life.

The transhuman project is based on the inscription of deterministic automatisms into cognitive activity. Assuming that conscious behavior is the effect of a deterministic chain of causation, the transhuman project purports to implement a technical automaton acting as a perfect replica of the human being: the android.

This project implies the insertion of deterministic chains into the epigenetic process itself; it aims at narrowing the space of epigenesis and subjecting epigenetic events to the determination of algorithmic chains inscribed in the body and mind. But the transhumanist project is based on a flawed idea of human experience.

Cognitive activity can be reduced to formal procedures, which in turn can be translated into algorithms, such that cognition can be replicated by artifacts and experience can become standardized. But the human agent cannot be reduced to his or her behavior and cognitive performances.

Although cognition and life can be simulated by intelligent automata, they are not reducible to the combination and embedding of information into artifacts. As highly complex and refined as they may be, constructs can develop intelligent behavior, but not experience. This is the philosophical flaw of transhuman ideology. And yet the development of cog-automation technology is nonetheless producing mutations in conscious organisms and in the social link. Transhuman ideology should not to be seen as a description, but as a project and a strategy to reprogram the human brain.

The Neuroplastic Dilemma

Like the concept of determinism, the concept of neuroplasticity is also double-edged: it describes the neural system as essentially plastic, but it also provides the conditions for executing a strategy. The plasticity of the neural system enables the project of neuro-subjection and uniformed cognitive mutation, such as the Google strategy. But on the other hand, the plasticity of the neural system implies that it is possible to develop a project of neuro-emancipation from surrounding reality.

A dilemma is overshadowing the near future: the dilemma between the adaptation of the neural system to a social and physical environment that is becoming increasingly intolerable for human sensibility—and the autonomous re-organization of the general intellect. The first case is already largely displayed in current social behavior: the globalized media-system exposes us daily to visions of unspeakable violence, massive torture, humiliation, misery, and the displacement and deportation of millions of women, children, and elderly people, but we are growing used to deactivating compassion, so much so that mass-cynicism now acts as a sort of ethical anesthesia.

The permanent exposure to horror deactivates ethical sentiment, acting as psychological habituation. Empathy is atrophied as an effect of this permanent exposition to horror. Neuroplasticity can thus imply a kind of apathetic and an-empathic automation of cognitive behavior, detached and *scotomized* from the emotional brain.

But the concept of neuroplasticity can be played in the opposite direction. The plasticity of the neural system can provide the condition for a fundamental reactivation of the psycho-cognitive apparatus in its social expression. Neuroplasticity can reactivate emotional empathy and political solidarity, necessary conditions for a process of self-organization of the general intellect driven by ethical and aesthetic sensibility rather than by the unethical impulse of economic competition.

The SCEPSI Experiment

In 2011, along with a group of artists and activists, I created the European School for Social Imagination (SCEPSI), aimed at creating a platform for the autonomy of the general intellect. The project of the SCEPSI intends to convert the social uprising into a process to disentangle the networked activity of scientists, engineers, and artists from the political and epistemological domination of the profit-oriented economy.

The SCEPSI has organized three international conferences, in San Marino, Barcelona and Kassel, which centered on the relation between financial power and knowledge, and the place of art in the self-organization of cognitive work.

Following the financial collapse of September 2008, European governments began a program of austerity based on the de-financing of social welfare, and of the educational system in particular. The downsizing and impoverishment of the public education system, resulting in the drastic reduction of resources, and the impoverishment and humiliation of teachers and researchers, paved the way for privatization. Students have been forced to take on debt to pay for their studies, while precariousness and unemployment have reached unprecedented levels. In the year 2011, the widespread Occupy movement emerged in many countries of the world: young precarious workers, students, and researchers launched a wave of protests against the de-financing of the public education system, and the privatization of knowledge, research, and training, occupying streets and squares, not universities.

At the time, I expected that a wave of occupations would take hold of European universities, and that from this long-lasting occupation the process of self-organizing knowledge would start. I was wrong. I had been reading the situation through the old categories of 1968. Students have completely changed, as has the university. Students are no longer people who spend their time studying, talking, and living together. They are forced to work and to bend to the blackmail of precarity. The university has become an empty place, where students only go to do their jobs, take their exams, and perform their competition. Relevant research is almost totally privatized.

Unable to endure over time and to begin a wave of long-lasting occupations of universities, the Occupy movement vanished, overpowered by two opposing and complementary forces: financial abstraction—the global chain of techno-linguistic automatisms—and identitarian aggressiveness—nationalism, fundamentalism, and fascism.

Occupy was an exceptional process of reactivating the social body, but it proved unable to turn the occupation of public spaces into a long-lasting process of social recomposition. Take, for example, the Egyptian catastrophe, the Syrian tragedy, or the wave of cynicism and depression in London after the four nights of rage in August 2011. The Occupy wave vanished, leaving behind a feeling of impotence and despair.

The attempt to instigate the self-organization of cognitive workers via a process of social uprising failed because precariousness and globalization have jeopardized the social solidarity that is necessary for a long-lasting process of autonomous organization. The history of movements for social emancipation has reached a turning point.

The abstract totalitarianism of finance and the aggressive identitarianism of disempowered populations are now the masters of the future, and political action will not dispel this alternative because the expanding crowds of unemployed, precarious workers, migrants, and displaced and deported people will be swallowed up in the identitarian whirlpool.

Neuro-Engineering: Consciousness and Evolution

Subsumption is interminable because of the unbridgeable gap between zero-dimensional and a-temporal information and the body as multidimensional and evolving-in-time. The game is over, yet the game is continuously renewed.

Neuroplasticity can pave the way to the brain's adaptation to an environment that is increasingly intolerable for a psychological, aesthetic, and ethical mind forged by the history of human civilization. Neo-human adaptation, adaptation to the connective mode of communication, to the ferocity of competition, to the barbarity and horror of the submission of life and attention to financial abstraction could take the form of a kind of social lobotomy: the pharmacological or surgical cancellation of what, in human psychology, is incompatible with the domination of abstraction.

Yet an alternative possibility lies in the brain's conscious ability for *self-plasmation*. This implies an autonomous recomposition of the living forces of the general intellect—a process of social organization of cognitive workers—and the recomposition of the social and erotic body of the general intellect.

In order to conceptualize the shift from past forms of political action, now devoid of effectiveness, to the evolutionary horizon of conscious neural evolution, a preliminary question must be answered: What is the relation between consciousness and evolution? Can we envisage an intentional activity aimed at a non-deterministic adaptation of the brain to the evolution of the environment? Can we imagine a conscious activity to orient the evolution of the brain? Can we consciously govern neural evolution?

In order to answer these questions, we should first consider the relation between aesthetic sensibility and the epistemic foundation of social action. Then we can focus on the creation of social, cultural, institutional, artistic, and neuro-engineering platforms for the self-organization of the general intellect, and for the recomposition of the networked activity of millions of cognitive workers worldwide with their social, erotic, and poetic bodies.

11

The End

La Malinche

Invoking La Malinche is the best way to speak of the end.

Humans have already experienced an end of the world, or the end of a world. A world ends when signs proceeding from the semiotic meta-machine grow undecipherable for a cultural community that perceived itself as a world.

A world is the projection of meaningful patterns on the surrounding space of lived experience. It is the sharing of a common code whose key lies in the forms of life of the community itself.

When flows of incomprehensible enunciations proceeding from the meta-machine invade the space of symbolic exchange, a world collapses because its inhabitants are unable to say anything effective about the events and things that surround them.

When signs proceeding from the environment are no longer consistent and understandable within the frame of the shared code, when the signs that convey effectiveness and potency escape the shared cultural code, a civilization ceases to be vital. It enters a tunnel of despair, quickly decays, and then dissolves. Its members die, or lose the ability to feel that they are part of a common, evolving reality, and those who survive undergo a process of integration into the code of an emerging culture, assimilating the colonizer's language and system of value.

From the point of view of the various indigenous cultures of pre-Columbian Mesoamerica, the Spanish colonization appeared as an end of the world, as the end of a world.

The Spaniards defeated the indigenous population thanks to their overwhelming military force, but colonization was essentially a process of symbolic and cultural submission. The "superiority" of the colonizers lay essentially in the operational effectiveness of their technical productions and expressions. Colonization destroyed the cultural environment in which indigenous communities had been living for centuries. Alphabetic technology and the power of the written word overwhelmed, jeopardized, and finally superseded indigenous cultures. The Christian message blended with the mythologies in existence prior to colonization, and modern Mexican culture emerged as an effect of the submission to alphabetic semiosis, as well as through contamination and syncretism.

The alphabetic meta-machine is based on the externalization of memory, and on the possibility of transferring information through time and space. Thanks to the functional superiority of their semiotic machine, the Europeans subjugated, subsumed, and re-coded the cultural universe of the natives, both in Mexico and in other areas of the continent.

Something similar is happening to us today. And so we must try to answer the question: What happens when a world dies, when outside flows of semiosis overpower and outperform existing languages and forms of life, and an entire world of values, expectations, and moral codes disintegrates?

At the bottom of the Latin American unconscious lies the myth of La Malinche.

Before the arrival of the Spanish invaders, Malinche (*Malinalli* in the Nahuatl language, or *Marina* for the Spaniards), the daughter of a noble Aztec family, was given away as a slave to passing traders after her father died and her mother remarried.

By the time Cortés arrived, she had learned the Mayan dialects spoken in the Yucatan while still retaining her knowledge of Nahuatl, the language of the Aztecs. For many years La Malinche—a resourceful woman of exceptional beauty and intellectual brilliance—became Cortés' lover and accompanied him as his interpreter. She translated exchanges between Cortés and Moctezuma, king of the Aztec population of Tenochtitlan, and translated the conqueror's words before crowds of indigenous people. She also translated the words of Christian conquerors and priests for the Nahuatl people.

In the novel *Malinche*, Laura Esquivel imagines Malinalli trapped between her own beliefs—taught in folktales and vivid imagery by her loving grandmother—and the Christian beliefs introduced by her master and lover.[1] How did she manage to translate Christian mythology and ethical concepts into the mythology of Quetzalcoatl and Huizilopochtli? What kind of symbolic transformation and re-elaboration did her translations involve? From the beginning of the conquest, in Mexico and in South America in general, Christian culture and mythology was reshaped in a syncretic way, and ambiguity was accepted as an essential feature of religious exchange. La Malinche the translator was doubly a traitor. Not only did she betray her own people, creating a link with the invaders, but she also betrayed the conquerors, and betrayed her lover himself. I use the word *betrayal* here only in a technical sense: from the moral point of view Malinche owed nothing to her own people who had sold her into slavery, and treated her as a servant. Cortés chose her as his lover and collaborator, and they had a child, Martin, who was the first Mexican. Malinche was extremely helpful to Cortés in his conquest. In a letter preserved in the Spanish archives, Cortés proclaimed that, "After God, we owe this conquest of New Spain to Doña Marina." Malinche's legacy is controversial: in contemporary Mexico, the word *malinche* is sometimes used pejoratively to describe someone who denies their heritage, who values other cultures above their own. Although she has been described as a traitor, historians contest that when conflict exploded between the Spanish and indigenous populations, Malinche played a key role in avoiding bloodshed. Her activity as a translator gave her the power to control information, and, most importantly, to translate concepts. Octavio Paz mentions Malinche in *The Labyrinth of Solitude*.

In contrast to Guadalupe, who is the Virgin Mother, the *Chingada* is the violated Mother. [...] Guadalupe is pure receptivity, and the benefits she bestows are of the same order: she consoles, quiets, dries tears, calms passions. The *Chingada* is even more passive. Her passivity is abject: she does not resist violence, but is an inert heap of bones, blood and dust. Her taint is constitutional and resides, as we said earlier, in her sex. This passivity, open to the outside world, causes her to lose her identity: she is the *Chingada*. She loses her name; she is no one; she disappears into nothingness; she *is* Nothingness. And yet she is the cruel incarnation of the feminine condition.

If the *Chingada* is a representation of the violated Mother, it is appropriate to associate her with the Conquest, which was also a violation, not only in the historical sense but also in the very flesh of Indian women. The symbol of this violation is doña Malinche, the mistress of Cortés. It is true that she gave herself voluntarily to the conquistador, but he forgot her as soon as her usefulness was over. Doña Marina becomes a figure representing the Indian women who were fascinated, violated or seduced by the Spaniards. And as a small boy will not forgive his mother if she abandons him to search for his father, the Mexican people have not forgiven La Malinche for her betrayal.[2]

Malinche is not only the expression of the mixing of cultures, but also the expression of the rebirth of the world from the collapse of the old. She is considered a symbol of subjection as well as a symbol of the emergence of a new Mexico, of a new history, and of a new world. But foremost, she is the expression of the consciousness that her world is over: she knows that her world as a system of consistent cultural and semiotic references has disintegrated.

If the limits of a world are the limits of the language that makes this world consistent and meaningful, Malinche is the symbol of the end of a world, and also the symbol of the formation of a new semiotic space of world-projection at the intersection of two different codes. Malinche is able to transform the collapse of her world into the creation of a new language, and therefore of a new world that is neither the prosecution of the old, nor the mere translation of the world of the conquistadores.

Only when one is able to see collapse as the obliteration of memory, identity, and as the end of a world can a new world be imagined. This is the lesson we must learn from Malinche.

The Cognitive Automaton and Us

Today, at the beginning of the twenty-first century, we are in a position that is similar to Malinche's: the conqueror is here, peaceful or aggressive, infinitely superior, unattainable, incomprehensible. We gave birth to this conqueror, this neo-human culture that emerged from our history and travelled beyond the ocean, destroying all forms of existing life to create a new code based on purity, and to engender the automaton, the rationale for unending automation.

The bio-info automaton takes shape at the point of connection between electronic machines, digital languages, and minds formatted in ways that comply with their codes. The automaton's flow of enunciation is emanating a connective world that conjunctive codes cannot interpret, a world that is semiotically incompatible with the social civilization that was the outcome of five centuries of humanism, Enlightenment thought, and socialism.

I will never be able to live in peace with the automaton, because I was formatted in the old world. As Pris says in *Blade Runner*, I'll die because I'm stupid. My body survives because I cannot find the way out. The human race is becoming an army of sleepwalkers: people suffering from Alzheimer's disease, taking pills, standing and facing reality, smiling, saying yes, yes, yes…

The automaton is the reification of the networked cognitive activity of millions of semio-workers around the globe. Only when they become compatible with the connective code can semio-workers enter the networking process. This implies the de-activation of conjunctive modes of communication and perception (compassion, empathy, solidarity, ambiguity and irony), paving the way to the assimilation of the conscious organism and the digital automaton.

According to transhumanist ideology, in a few decades, digital automatons will be able to perfectly replace human organisms. Ray Kurzweil, for instance, thinks that in the near future humans and machines will become interchangeable from the point of view of cognitive efficacy. This is clearly possible. Yet the implication that the automaton and the human will merge is false, for the automaton can never be assimilated to the human because the specificity of the human lies in the relation between conscious rationality and the unconscious.

The automaton's functional cognition is more powerful than human cognition from the operational point of view. It is more powerful, more effective, and obviously more destructive. But the unbridgeable difference between the conscious organism and the automaton—as complex and refined as it may be—lies in the unconscious.

The automaton's unconscious is the material hardware of the electro-magnetic machinery that we call the Net. The human unconscious, on the other hand, is fleshy, marked by ambiguity, inconsequentiality, and, most importantly, by death.

The automaton is pure functionality, even when it is endowed with self-regulating evolution. It will subsume human cognitive competence and subject it to its rule. So the prospect that we will have to face is not the sweetish transhuman alliance between friendly, hyper-intelligent machines and human beings; rather, it is the final subjection of humans to the rule of non-organic intelligent automata whose behavior will be regulated according to criteria inscribed in them by their maker, bio-financial capitalism.

Surely the automaton will be able to evolve. But its maker will have inscribed the paradigm of this evolution in its info-genetic code. And the maker coincides with the most advanced corporations of bio-financial capitalism such as Google.

In today's global landscape, after the disappearance of egalitarian cultures, there are but two actors: the first is the all-pervading force of financial abstraction, the second, the proliferation of rancorous, reactive, identitarian bodies.

Financial abstraction is based on the faceless operativity of automatisms embedded in soulless social dynamics. Nobody is really in charge; nobody is making conscious decisions. In economic operations, logical mathematical implications have replaced deciders, and the algorithm of capital has grown independent of the individual wills of its owners.

The impersonality of financial abstraction escapes any attempt at conscious political transformation, so that people who have lost control of their life hang on dearly to a sense of illusory belonging. The nation, religious faith, and ethnicity provide protection from insecurity and loneliness, and serve as tools to attack competitors.

The connective energies of the new generation have been recombined by the techno-financial automaton, and reduced to a condition of precariousness. Aggressive belonging is their only form of cohesion.

Will the general intellect be able to disentangle itself from the automaton? Can consciousness act on neurological evolution? Will humans be able to find a new conjunctive language, in the connective kingdom of digital code? Will pleasure, affection, and empathy find ways to re-emerge out of their conjunctive framework? Will we be able to translate into human language the connective language of the automated semio-machine whose buzzing is growing in our heads?

These are questions that only Malinche can attempt to answer, opening to the incomprehensible other, betraying her people, and reinventing language in order to express what cannot be said.

—August 2014

Notes

Introduction: Concatenation, Conjunction, and Connection

1. The word geo-culture was proposed by Irit Rogoff in "Geo-cultures, circuits of art and globalization," *Open 16: The Art Biennial as a Global Phenomenon: Strategies in Neo-Political Times* (2009).

2. Henri Bergson, *Creative Evolution*, trans. Arthur Mitchell (London, New York: MacMillan, 1911), 11.

3. "I could never know to what degree I was the perpetrator, configuring the configurations around me, oh, the criminal keeps returning to the scene of the crime! When one considers what a great number of sounds, forms reach us at every moment of our existence ... the swarm, the roar, the river ... nothing is easier than to configure! Configure! For a split second this word took me by surprise like a wild beast in a dark forest, but it soon sank into the hurly-burly of the seven people sitting here, talking, eating, supper going on [...]" Witold Gombrowicz, *Cosmos*, trans. Danuta Borchardt (New Haven: Yale University Press, 2005), 54–55.

4. See Gregory Bateson, *Mind and Nature: A Necessary Unity* (New York: E.P. Dutton, 1979).

5. See Paolo Virno, *Saggio sulla negazione: per una antropologia linguistica* (Turin: Bollati Boringhieri, 2013) (*Essay on Negation: Towards a Linguistic Anthropology*, untranslated).

6. See William Burroughs, *Ah Pook Is Here* (New York: Riverrun Press, 1979).

7. See Simon Baron-Cohen, *Zero Degrees of Empathy: A New Theory of Human Cruelty* (New York: Penguin, 2011).

8. "'They do say war is a bit like playing chess.'
'Yes it is,' said Prince Andrey, 'but there's one little difference. In chess you can take as long as you want over each move. You're beyond the limits of time. Oh, there is this one other difference: a knight is always stronger than a pawn, and two pawns are always stronger than one, whereas in war a battalion can sometimes be stronger than a division, and sometimes weaker than a company. You can never be sure of the relative strength of different forces. Believe me,' he went on, 'if anything really depended on what gets done at headquarters, I'd be up there with them, doing

things, but no, I have the honor of serving here in this regiment along with these gentlemen, and I'm convinced that tomorrow's outcome depends on us, not on them ... Success never has depended, never will depend, on dispositions or armaments, not even numbers, and position least of all.'

'Well, what does it depend on?'

'On the gut feeling inside me and him,' he indicated Timokhin, 'and every soldier.'" Leo Tolstoy, *War and Peace*, trans. Anthony Briggs (London: Penguin Classics, 2005), 858.

9. Hubert L. Dreyfus, *What Computers Can't Do* (New York: Harper Collins, 1979), 33.

10. *Ibid.*, 68–69.

11. *Ibid.*, 71.

1. The Sensitive Infosphere

1. Gilles Deleuze, *Francis Bacon: The Logic of Sensation*, trans. Daniel W. Smith (Continuum, London, 2003), 45. Originally published in France in 1981.

2. Deleuze and Guattari, *A Thousand Plateaus* (Minneapolis, London: University of Minnesota Press, 1987),153.

3. In Guattari's parlance, a *ritornello* is a recurring semiotic concatenation that links a subject with the surrounding environment and the cosmos. A song, a slogan, a ritual, a symbol, a viral message may be seen as refrains.

4. Gabrielle Dufour-Kowalska, *L'art et la sensibilité de Kant à Michel Henry* (Paris: Vrin, 1996), 11–12. My translation.

5. See Michel Foucault, *The Order of Things: An Archeology of the Human Sciences* (New York: Pantheon Books, 1970). Originally published in France in 1966.

6. Jonathan Crary, *24/7: Late Capitalism and the Ends of Sleep* (London: Verso, 2013), 40.

7. Max Pagès, *Trace ou sens: le système émotionnel* (Paris: Hommes et groupes, 1986), 109.

8. *Ibid.*, 118.

2. Global Skin: A Trans-Identitarian Patchwork

1. David Hellerstein, "Skin," *Science Digest* (September 1985).

2. Daniel Heller-Roazen, *The Inner Touch: Archeology of a Sensation* (New York: Zone Books, 2007), 17.

3. Diane Ackermann, *A Natural History of the Senses* (New York: Vintage Books, 1990), 95.

4. Heller-Roazen, *The Inner Touch*, 81.

5. Irit Rogoff, *Terra Infirma: Geography's Visual Culture* (New York: Routledge, 2000), 15.

6. V.S. Naipaul, *The Enigma of Arrival* (New York: Viking, 1987), 144.

7. V.S. Naipaul, *An Area of Darkness* (London: Deutsch, 1964; New York: Vintage, 2002), 290. Citations refer to the Vintage Edition.

8. V.S. Naipaul, *The Overcrowded Barracoon and Other Articles* (London: Deutsch, 1972), 24–5.

9. V.S. Naipaul, *India: A Wounded Civilization* (London: Deutsch, 1977; New York, Vintage, 2003), 10. Citations refer to the Vintage Edition.

10. V.S. Naipaul, *Beyond Belief: Islamic Excursions among the Converted Peoples* (London: Random House, 1998), 58.

11. Fatima Mernissi, *Beyond the Veil* (Bloomington: Indiana University Press, 1987), 17.

12. Ibn Hazm, *The Ring of the Dove*, trans. A.J. Arberry (London: Luzac and Company, 2004). Available online at http://www.muslimphilosophy.com/hazm/dove/chp25-26.html.

13. Denis de Rougemont, *Love in the Western World* (New York: Panthon Books, 1940; Princeton, NJ: Princeton University Press, 1983), 50. Citations refer to the Princeton U.P. Edition.

14. Denis de Rougemont, *Love Declared: Essays on the Myths of Love*, trans. Richard Howard (New York: Pantheon Books, 1963), 41–42.

15. Guido Guinzinelli, "Of the gentle Heart," in *Dante and His Circle: With the Italian Poets Preceding Him,* trans. Dante Gabriel Rossetti (Boston: Roberts Brothers, 1887), 187–189.

16. Denis de Rougemont, *Love Declared*, 41.

17. Charles Baudelaire, "Hymn to Beauty," *The Flowers of Evil*, trans. James McGowan (Oxford: Oxford University Press, 1993), 45.

18. Stéphane Mallarmé, "Sigh," *Collected Poems*, trans. Henry Weinfield (Berkeley: University of California Press, 1994), 22.

19. "Sea Breeze," *Ibid*, 21.

20. See Mario Praz, *Romantic Agony*, trans. Angus Davidson (London and New York: Oxford University Press, 1933).

21. Natalia Ilyin, *Chasing the Perfect: Thoughts on Modernist Design in Our Time* (New York: Bellerophon Publications, 2006), 31.

22. Wilhelm Worringer, *Abstraction and Empathy: A Contribution to the Psychology of Style*, trans. Michael Bullock (Chicago: Elephant Publishers, 1997), 4.

23. See Sigmund Freud, "Obsessive Actions and Religious Practice," *The Freud Reader* (New York: W.W. Norton & Company, 1989).

24. Nikolaj Alexandrovic Berdyaev, "The Revelation about Man in the Creativity of Dostoevsky," trans. Fr. S. Janos, available online at http://www.berdyaev.com/berdiaev/berd_lib/1918_294.html.

25. Vittorio Strada, *La questione russa: identità e destino* (Venezia: Marsilio, 1991).

26. Cited in Benedikt Sarnov, *Nash sovetskii novoiaz: malen'kaia entsiklopediia real'nogo sotsializma* (Moscow: Materik, 2002), 446–447.

27. See Hélène Carrère d'Encausse, *Lénine: la révolution et le pouvoir* (Paris: Flammarion, 1979).

28. In the Italian jargon of *operaismo*, we say *estraneità*, which means being outside, being external, extraneous, and therefore autonomous.

29. Berdyaev, "The New Russia," trans. by Fr. S. Janos, available online at http://www.berdyaev.com/berdiaev/berd_lib/1915_188.html.

30. See Takeo Doi, *The Anatomy of Dependence: The Key Analysis of Japanese Behavior*, trans. John Bester (Tokyo: Kodansha International, 1973)

31. See Oguma Eiji, *A Genealogy of Japanese Self-Images*, trans. David Askew (Melbourne: Trans Pacific Press, 1995).

32. Yosiyuki Noda, *Introduction to Japanese Law*, trans. and ed. Anthony H. Angelo (Tokyo: University of Tokyo Press, 1976), 170.

33. Natsume Soseki, *Theory of Literature and Other Critical Writings* (New York: Columbia University Press, 2009), 48.

34. Petrarch, *The Canzoniere, or Rerum vulgarium fragmenta*, trans. Mark Musa (Bloomington, IN: Indiana University Press, 1996), 205.

35. From *tenno*, the Japanese term for emperor.

36. Yoshikuni Igarashi, *Bodies of Memory: Narratives of War in Postwar Japanese Culture, 1945–1970* (Princeton: Princeton University Press, 2000), 28–29.

37. Quoted in *ibid*, 29.

38. Adrian Favell, *Before and after Superflat: A Short History of Japanese Contemporary Art, 1990–2011* (Hong Kong: Blue Kingfisher, 2011), 19.

39. *Ibid.*, 40.

40. *Ibid.*, 11.

41. *Ibid.*, 31.

42. Robert Koehler et al., *Hangeul: Korea's Unique Alphabet* (Seoul: Seoul Selection, 2010), 60.

3. The Aesthetic Genealogy of Globalization

1. See for example Geert Lovink, Derrick De Kerkhove, and Manuel Castells, among others.

2. See Gilles Deleuze, *The Fold: Leibniz and the Baroque*, trans. Tom Conley (London: the Athlone Press, 1993).

3. See José Antonio Maravall, *Culture of the Baroque: Analysis of a Historical Structure* (Manchester: Manchester University Press, 1986).

4. See Néstor Luján, *La vida cotidiana en el Siglo de Oro* (Barcelona: Planeta, 1988).

5. See Francisco de Quevedo and Anonymous, *Lazarillo de Tormes and The Swindler (El Buscón): Two Spanish Picaresque Novels*, trans. Michael Alpert (London and New York: Penguin Books, 2003).

6. Maravall, *Culture of the Baroque*, 174–193.

7. Frederick Jackson Turner, *The Significance of the Frontier in American History* (New York: Penguin Books, 2008), 48. First published in 1893.

8. Carl Degler, *Out of Our Past* (New York: Harper and Row, 1959), 16.

9. Camille Paglia, *Sex, Art, and American Culture* (New York: Vintage, 1992), 29.

10. Jean Baudrillard, *Symbolic Exchange and Death*, trans. Iain Hamilton Grant (London: Sage Publications, 1993), 2.

11. Exodus, 20:2–6.

12. Alain Besançon, *The Forbidden Image: An Intellectual History of Iconoclasm* (Chicago: Chicago University Press, 2000), 79.

13. Quoted in John Renard, *Seven Doors to Islam: Spirituality and the Religious Life of Muslims* (Berkeley: University of California Press, 1996), 126.

14. Shaker Laibi, *Soufisme et art visuel: iconographie du sacré* (Paris: L'Harmattan, 1998), 14. My translation.

15. Dominique Clévenot, *Une esthétique du voile: essai sur l'art arabo-islamique* (Paris: L'Harmattan, 1994), 85. My translation.

16. *Ibid.*, 8.

17. *Sadih al-Bukhari*, quoted in ibid., 86.

18. Fethi Benslama, *Psychoanalysis and the Challenge of Islam*, trans. Robert Bononno (Minneapolis: University of Minnesota Press, 2009), 136.

19. *Ibid.*, 138–139.

20. Marie-José Mondzain, *Image, Icon, Economy: The Byzantine Origins of the Contemporary Imagination*, trans. Rico Franses (Stanford: Stanford University Press, 2004).

21. Nicholas Mirzoeff, *An Introduction to Visual Culture* (New York: Routledge, 2009), 5.

22. Serge Gruzinsk, *Images at War: Mexico From Columbus to Blade Runner (1492–2019)*, trans. Heather MacLean (Durham: Duke University Press, 2001).

Part 2. The Body of the General Intellect

1. Karl Marx, *Grundrisse*, trans. Martin Nicolaus (London: Penguin, 1973), 706.

4. Language, Limit, Excess

1. Quoted in Martin Heidegger, *Poetry, Language, Thought* (New York: Harper-Collins, 1971), 216.

2. Lama Anagarika Govinda, *Foundations of Tibetan Mysticism* (Boston, MA: Weiser Books, 1969), 22.

3. From Blok's diary entry, June 29, 1909.

4. Raghava Menon, *Indian Classic Music: An Initiation* (Delhi: Vision books, 1997), 58.

5. See Félix Guattari, *The Machinic Unconscious: Essays in Schizoanalysis* (Los Angeles: Semiotext(e), 2001).

6. Mircea Eliade, *Images and Symbols: Studies in Religious Symbolism*, trans. Philip Mairet (Princeton: Princeton University Press, 2001), 20.

7. Govinda, *Foundations of Tibetan Mysticism*, 19.

8. *Ibid.*, 19.

9. Stéphane Mallarmé, *Selected Poetry and Prose*, trans. Mary Ann Caws (New York: New Directions, 1982), 75–76.

10. Giorgio Agamben, *Language and Death* (Minneapolis: University of Minnesota Press, 2006).

11. Robert Sordello, *Money and the Soul of the World* (Dallas: The Pegasus Foundation, 1983).

12. Marshall McLuhan, *Understanding Media: The Extensions of Man* (Berkeley: Ginko Press, 1964; Cambridge, MA: MIT Press, 1994), 136. Citations refer to the MIT Press Edition.

13. Jean Baudrillard, *The Mirror of Production*, trans. Mark Poster (Saint Louis: Telos Press 1975), 30.

14. McLuhan, *Understanding Media*, 187.

15. *Ibid.*, 191.

16. Jean Baudrillard, *Symbolic Exchange and Death*, 3.

17. *Ibid.*, 6–7.

18. See Arthur Kroker, *Data Trash: The Theory of the Virtual Class* (New York: Saint Martin's Press, 1994).

19. Arthur Rimbaud, "Letter to George Izambard, May 1871," *Selected Poems and Letters*, trans. John Sturrock (New York: Penguin Classics, 2005), 236.

20. Baudrillard, *Symbolic Exchange and Death*, 2–9.

21. Deleuze and Guattari, *A Thousand Plateaus*, 7–10.

22. *Ibid.*, 10.

5. Avatars of the General Intellect

1. Marx, *Grundrisse*, 704–706.

2. *Ibid.*

3. *Ibid.*

4. *Ibid.*, 138.

5. *Ibid.*, 142.

6. *Ibid.*

7. Mario Savio, Berkeley 1964. Quoted in Hal Draper, *Berkeley: The New Student Revolt* (Grove Press, New York, 1965).

8. Fred Turner, *From Counterculture to Cyberculture* (Chicago: University of Chicago Press, 2006), 13.

9. John Perry Barlow, "A Cyberspace Independence Declaration," http://w2.eff.org/Censorship/Internet_censorship_bills/barlow_0296.declaration.

10. Turner, *From Counterculture to Cyberculture*, 22.

11. See Robert Jungk, *Brighter than a Thousand Suns: A Personal History of the Atomic Scientists*, trans. Victor Gollancz (New York: Harcourt Brace, 1958).

12. McLuhan, *Understanding Media*, 3.

13. Esther Dyson, George Gilder, George Keyworth, and Alvin Toffler, "Cyberspace and the American Dream: A Magna Carta for the Knowledge Age," *The Freedom*

and Progress Foundation online, http://www.pff.org/issues-pubs/futureinsights/fil.2magnacarta.html.

14. Kevin Kelly, *Out of Control: The New Biology of Machines, Social Systems & the Economic World* (Boston: Addison-Wesley Publishing Co., 1994), 33.

15. *Ibid.*, 36.

6. The Swarm Effect

1. Kelly, *Out of Control*, 2.

2. Kelly, *Out of Control*, 202.

3. Eugene Thacker, "Networks, Swarms, Multitudes, Part Two," *CTheory*, http://www.ctheory.net/articles.aspx?id=423.

4. Ross Douthat, "The Great Consolidation," *The New York Times*, May 16, 2010. http://www.nytimes.com/2010/05/17/opinion/17douthat.html?_r=0.

7. Social Morphogenesis and Neuroplasticity

1. Oliver Sacks, *Migraine* (New York: Vintage, 1999), 34.

2. Catherine Malabou, *The New Wounded: From Neurosis to Brain Damage*, trans. Steven Miller (New York: Fordham University Press, 2012), 4.

3. *Ibid.*, 35.

4. *Ibid.*, 9.

5. Dimitris Papadopoulos, "The Imaginary of Plasticity: Neural Embodiement, Epigenetics and Ecomorphs," *The Sociological Review*, 59:3 (2011): 433.

6. *Ibid.*, 439–440.

7. Catherine Malabou, *What Should We Do With Our Brain?* (New York: Fordham University Press, 2008), 80.

8. Antonio Damasio, *The Feeling of What Happens: Body and Emotions in the Making of Consciousness* (New York: Harcourt, Brace and co. 1999), 313.

9. Luisa Muraro, *L'ordine simbolico della madre*, (Roma: Editori Riuniti, 1991).

10. Kalle Lasn, *Meme Wars: The Creative Destruction of Neoclassical Economics* (New York: Seven Stories Press, 2012).

11. *The Psychopathologies of Cognitive Capitalism, Part One,* hosted by Arne de Boer, Warren Niedich, and Jason Smith, was organized at CalArts in Los Angeles, with speakers including myself, Jonathan Beller, Jodi Dean, Tiziana Terranova, Patricia Pisters, and Bruce Wexler. The second part of the conference was held at the ICI Berlin in March 2013.

12. See in particular Papadopoulos, "The Imaginary of Plasticity."

13. Félix Guattari, "Machine and Structure," *Molecular Revolution: Psychiatry and Politics*, trans. Rosemary Sheed (New York: Penguin, 1984), 111–19.

14. *Ibid.*

15. Félix Guattari, *Chaosmosis* (Bloomington & Indianapolis: Indiana University Press, 1995), 135.

16. Deleuze and Guattari, *What Is Philosophy?*, trans. Hugh Tomlinson and Graham Burchell III (New York: Columbia University Press, 1996), 201.

17. Guattari, *Chaosmosis*, 128.

18. McLuhan, *Understanding Media*, 5–8.

19. Alvin Toffler, *The Third Wave* (New York: William Morrow and Company, 1980), 10.

20. François Lyotard, *The Postmodern Condition; A Report on Knowledge*, trans. Geoff Bennington and Brian Massumi (Minneapolis: University of Minnesota Press, 1984), 3.

21. Gregory Bateson, *Steps to an Ecology of Mind* (New York: Ballantine books, 1972; Chicago: University of Chicago Press, 2000), 468. Citations refer to the Chicago Press edition.

22. *Ibid.*, 261.

8. The Transhuman

1. Jean-François Lyotard, *The Inhuman* (Stanford: Stanford University Press, 1991), 118.

2. Martin Heidegger, "Letter on 'Humanism,'" *Pathmarks*, trans. Frank A Capuzzi (Cambridge: Cambridge University Press, 1998), 239, 247, and 249.

3. Giovanni Pico della Mirandola, *Oration on the Dignity of Man* (Adelaide: The University of Adelaide Library, 2014), 6–7.

4. Martin Heidegger, *Off the Beaten Track*, trans. Julian Young and Kenneth Haynes (Cambridge, UK: Cambridge University Press, 2002), 71.

5. Christian Marazzi, "Measure and Finance," talk presented at the conference *Measure for Measure: A Workshop on Value from Below*, Goodenough College, London, September 21, 2007. The text is available online at: http://www.generation-online.org/c/fc_measure.htm.

6. Bill Joy, "Why the Future Doesn't Need Us," *Wired* 8, no 04 (April 2000), http://archive.wired.com/wired/archive/8.04/joy.html.

7. Ray Kurzweil, *The Singularity is Near: When Humans Transcend Biology* (New York: Viking Books, 2005), 163–181.

8. *Ibid.*, 162–164.

9. Thomas Metzinger, *The Ego Tunnel: The Science of Mind and the Myth of the Self* (New York: Basic books, 2009), 1–2.

10. *Ibid.*, 7.

11. In Stanislaw Ulam, "Tribute to John von Neumann," *Bulletin of the American Mathematical Society* 64, no 3, part 2 (May 1958): 5.

12. Henri Bergson, *Creative Evolution*, 40.

13. *Ibid.*, 39–40.

14. *Ibid.*, 11.

15. *Ibid.*, 48 and 17.

16. Douglas Hofstadter, *Gödel, Escher, Bach: An Eternal Golden Braid* (New York: Basic Books, 1979), 26.

17. *Ibid.*, 26–27.

18. *Ibid.*, 598.

19. See Heller-Roazen, *The Inner Touch*.

20. Metzinger, *The Ego Tunnel*, 15.

9. The Horizon of Mutation

1. Heidegger, *Off the Beaten Track*, 73.

2. Steven Shapiro, "Accelerationist Aesthetics: Necessary Inefficiency in Times of Real Subsumption," *e-flux journal* 46 (June 2013), http://www.e-flux.com/journal/accelerationist-aesthetics-necessary-inefficiency-in-times-of-real-subsumption/.

3. Sections of this chapter were previously published as a short pamphlet entitled *Neuro-totalitarianism in Technomaya: Goog-Colonization of Experience and the Neuro-Plastic Alternative* (Los Angeles: Semiotext(e), 2014).

4. Quoted in Mircea Eliade, *Images and Symbols*, 70–71.

5. Aldoux Huxley, *Texts and Pretexts* (London: Chato & Windus, 1932), 12.

6. Merleau-Ponty, *Phenomenology of Perception*, trans. Colin Smith (London: Routledge, 1962), XIX.

7. *Ibid.*, XVIII.

8. Edmund Husserl, *Experience and Judgment*, trans. J.S. Churchill and K. Ameriks (London: Routledge, 1975), 41.

9. *Ibid.*, 27.

10. Jonathan Crary, *24/7: Late Capitalism and the Ends of Sleep*, 40.

11. *Ibid.*, 11.

12. *Ibid.*, 10.

13. *Ibid.*, 74.

14. Gary Genosko, *When Technocultures Collide: Innovation from Below and the Struggle for Autonomy* (Waterloo, Canada: Laurier University Press, 2013), 149.

15. See Jean Piaget and Bärbel Inhelder, *The Child's Conception of Space: A Study of the Development of Imaginal Representation*, trans. F.J. Langdon and J.L. Lunzer (New York: W.W. Norton & Co., 1967).

16. Jorge Luis Borges, "In Praise of Darkness," in *Selected Poems 1923–1967* (London: Penguin, 1985), 299.

17. Franco La Cecla, *Perdersi*, (Bari: Laterza, 1988), 43. My translation.

18. *Ibid.*, 91. My translation.

19. Rainer Maria Rilke, *Letters of Rainer Maria Rilke: 1910–1926*, trans. Jane Banard Greene and M.D. Herter Norton (New York: W.W. Norton & Co., 1947), 374.

20. Jürgen Habermas, *The Theory of Communicative Action. Volume 2: Lifeworld and System: A Critique of Functionalist Reason*, trans. Thomas Mc Carthy (Boston: Beacon Press, 1987), 153.

21. Alessandro Sarti and Giovanna Citti, "On the Origin and Nature of Neurogeometry," *La Nuova Critica* 55–56 (2010).

22. Félix Guattari, *Chaosmosis*, 16.

23. Deleuze and Guattari, *What is Philosophy?*, 204.

11. The End

1. See Laura Esquivel, *Malinche*, trans. Ernesto Mestre-Reed (New York: Washington Square Press, 2006).

2. Octavio Paz, *The Labyrinth of Solitude: The Other Mexico, Return to the Labyrinth of Solitude, Mexico and the United States, the Philanthropic Ogre* (New York: Grove Press, 1985), 77.